A PLUME BOOK
THE BOOK OF ORIGINS

TREVOR HOMER is a former champion amateur golfer who represented England seventeen times. He has been a director of several private and public companies, and lives in Staffordshire, England. This is his first book, based on a lifelong obsession with obscure facts.

D0391445

THE BOOK OF
ORIGINS

DISCOVER THE AMAZING ORIGINS OF
THE CLOTHES WE WEAR, THE FOOD WE EAT,
THE PEOPLE WE KNOW, THE LANGUAGES
WE SPEAK, AND THE THINGS WE USE

TREVOR HOMER

A PLUME BOOK

PLUME
Published by Penguin Group
Penguin Group (USA) Inc., 375 Hudson Street, New York, New York 10014, USA • Penguin Group (Canada), 90 Eglinton Avenue East, Suite 700, Toronto, Ontario, Canada M4P 2Y3 (a division of Pearson Penguin Canada Inc.) • Penguin Books Ltd., 80 Strand, London WC2R 0RL, England • Penguin Ireland, 25 St. Stephen's Green, Dublin 2, Ireland (a division of Penguin Books Ltd.) • Penguin Group (Australia), 250 Camberwell Road, Camberwell, Victoria 3124, Australia (a division of Pearson Australia Group Pty. Ltd.) • Penguin Books India Pvt. Ltd., 11 Community Centre, Panchsheel Park, New Delhi – 110 017, India • Penguin Group (NZ), 67 Apollo Drive, Mairangi Bay, Auckland 1311, New Zealand (a division of Pearson New Zealand Ltd.) • Penguin Books (South Africa) (Pty.) Ltd., 24 Sturdee Avenue, Rosebank, Johannesburg 2196, South Africa

Penguin Books Ltd., Registered Offices: 80 Strand, London WC2R 0RL, England

Published by Plume, a member of Penguin Group (USA) Inc. Originally published in Great Britain in different form by Portrait, an imprint of Piatkus Books Limited.

First American Printing, June 2007
10 9 8 7

 REGISTERED TRADEMARK—MARCA REGISTRADA

LIBRARY OF CONGRESS CATALOGING-IN-PUBLICATION DATA

Homer, Trevor.
  The book of origins : discover the amazing origins of the clothes we wear, the food we eat, the people we know, the languages we speak, and the things we use / Trevor Homer.
   p. cm.
  Originally published: London: Portrait, 2006.
  Includes bibliographical references and index.
  ISBN 978-0-452-28832-4
  1. Handbooks, vade-mecums, etc. 2. Civilization—Miscellanea. I. Title.
  AG106.H66 2007
  031.02—dc22   2006033082

Printed in the United States of America
Designed by Virginia Norey

To my two exceptional sons,
Max and James.
This is for you.

CONTENTS

CONTENTS

ACKNOWLEDGMENTS

My thanks are due to the many hundreds of specialists who compile reference works. The *Encyclopaedia Britannica* has been my bible and Google is god. The British Library has been a second home for two years and the library of the Wellcome Trust, the greatest depository of medical knowledge in the world, has provided me with all I could ever need about the origins of medicine.

I thank John Connolly (in memoriam) who first said I should, Bill Taylor who said it second, and Keith Ward who said it third.

I have been fortunate to have Andrew Lownie as my agent. He gave me the unfailing encouragement and support that all writers need, and indeed crave.

My thanks also to Stephanie Hale, David Haviland, and Carl Cutler, who edited my manuscript with sympathy and consummate skill. They corrected mistakes and improved my English. Any errors remaining are entirely my own. Among the many who responded to my e-mails, special thanks to April Ashley and Herbert Deutsch who helped me with dates.

I am particularly indebted to Alan Brooke at Portrait, who kept me on track when I began to deviate.

For special mention, I would like to single out Emily Haynes of Plume. I am very grateful for all your help and for your painstaking work with the manuscript.

I would like to thank my dear sister Dianne who did the early reading. It couldn't have been easy.

I have been lucky that Susan, to whom I have been married for thirty years, never lost faith and allowed me to work late. I am aware of the sacrifices, and you were always there.

INTRODUCTION

The creation of a thousand forests is in a single acorn.
—RALPH WALDO EMERSON (1803-82)

Everything has an origin. This book is for people who want to know how and when things began, where they came from, and why they started. It celebrates the work of explorers, scientists, and inventors, pioneers who wanted to go further than anyone had gone before them—people who wanted to know what was over the next hill or beyond the ocean—people who wanted to know how the world works and ended up discovering or inventing something that no one had ever seen before.

Some things have an entirely unsuspected origin, for example the Hollywood "tough guy" career of James Cagney (see "Famous People," p. 99), and other origins have been wrongly attributed to famous people. Hedy Lamarr, the great beauty and Hollywood star of the 1930s, invented something that affects almost all of our everyday lives, and which could have brought her a major fortune. She earned not a penny from it (see "Communication," p. 28 and "War," p. 260).

Ancient cave painters began something that led to the sublime work of Michelangelo and Leonardo da Vinci. Everyday items such as the clothes we wear, the food we eat, the phones we use, the televisions we watch, the medicines that cure us, the sports we play, the languages we use, as well as capitalism, began somewhere.

From the simple hunting and gathering activities of the earliest ape-like creatures, through to the advances made by today's sophisticated human beings, the pace of development has been astonishing, and it all had an origin. We human beings, alone among the animals, changed cosmic dust into axes, motorcars, palaces, films, computers, perfumes, and sausages (see "Food and Drink," p. 113).

We even started religions, wars, and political parties. We human beings, without any innate ability to take to the air, have developed machines to fly to the other side of the world in the same time it used to take our grandparents to cross a couple of counties. Human beings invented a device that enables a person, at the touch of just a few buttons, to speak to another person in the farthest places of the world. Human beings watched on live television as men walked on the surface of the moon, and, most importantly, human beings routinely cure diseases that once killed millions (see "Health," p. 132).

We invented abstract thought, such as philosophy or budgeting. The signs for add, subtract, multiply, and divide, which are so much a part of our existence that we most likely never even think about them, began somewhere. We also invented question marks, poison gas, and money. None of it, not even the Beatles (originally the Quarrymen), came to us fully formed. Some of the things we look at, such as frozen peas or the second New Zealand—that is, the New Zealand that disappeared off the map in 1792 (see "Countries and Empires," p. 38)—have little to do with human creativity.

So, where has it all come from? What are the origins? The Chinese have an elegant proverb: "With time and patience, a mulberry leaf becomes a silk gown."

Looking at a silk gown, and knowing nothing about the digestive systems of silkworms or their diet, who could imagine such a beautiful object could owe its origin to a simple mulberry leaf?

From art to war, via sports, language, and buildings, with some crime and sex thrown in, let's take a look at origins.

ART

COVERING PAINTING, SCULPTURE, MUSIC, OPERA,
POETRY AND LITERATURE, AND DANCE

All art has this characteristic—it unites people.
—LEO TOLSTOY (1828-1910)

PAINTING

Most people associate painting and art so closely that they are almost interchangeable, so it is appropriate that painting is the first of the art forms.

CAVE PAINTING

The earliest of all art forms, cave painting dates from approximately 40,000 BC. An explosion of creativity started around 35,000 BC, lasting to about 14,000 BC until it ceased completely in 10,000 BC. This is known as the Upper Paleolithic period.

The western European cave artists are unknown by name, but belonged to the people referred to generally as the Magdalenians, after one of the sites, La Madeleine, in the Dordogne region of France. They mainly painted animals at a large scale, including now extinct species such as the woolly mammoth and the woolly rhinoceros. They also painted human hands and used signs and geometric shapes, but they did not depict ceremonial or sacrificial events. The main colors used were black and red, with some white, brown, and yellow.

The sheer scale of some of the paintings and the fact that they appear on inaccessible surfaces high up in the roofs of caves suggest that it is entirely possible the artists were professionals who were rewarded for their efforts rather than individuals who undertook the work for personal pleasure.

There is evidence from socket holes in the cave walls that scaffolding and platforms were required to execute some of the work. It is now believed that the artists could have been part of an organized studio system, which provided the decorations for the occupants.

The first discovery of cave paintings was made at Altamira, near Santander in northern Spain, in 1879 by Maria de Sautuola, a nine-year-old girl who was helping her father, a local archaeologist, to explore a cave system in search of ancient bones. Walking ahead of her father, she became the first person for more than 30,000 years to see what is regarded by experts as some of the finest cave art so far discovered. The roof of the main cave is covered with paintings of wild boar, bison, a deer, and some horses, executed in vivid red, violet, and black. The artist also left handprints and hand outlines, as if signing the work.

Australian cave art Cave paintings have also been discovered on the Arnhem Land peninsula in Australia, and these may be older than the European examples. Some sources speculate that Australian cave art may span the whole period of human habitation on Australia, some 60,000 years. Traditional subjects are still painted today by Aboriginal men.

British cave art Until recently it was thought that Britain had no cave art. However, in 2003, bas-reliefs and paintings were discovered at Church Hole cave on the Nottinghamshire–Derbyshire border. They were dated at 15,000 years old and are now regarded as being among the best examples of cave paintings in existence.

NATIVE AMERICAN CAVE ART

Only discovered in the late 1970s, the earliest known examples of Native American cave art are in the Adair Glyph cave, in Adair County, Kentucky. These have been dated as early as 1610 BC by carbon dating charcoal deposits on the cave floor, which were the result of burning cane torches.

These early Native American artists had begun to produce complex and well-structured drawings in soft mud, known as "mud glyphs," more than a half-mile from the cave entrance, well beyond the reach of natural light. They selected as their subject matter warriors, animals, and winged human beings.

In a secret and as yet unnamed cave in Tennessee, elaborate petroglyphs (marks scratched into the rock walls and roofs of the cave), have been provisionally dated at 4,300 years old.

Although we can never know for certain, it is thought that prehistoric tribes considered the cave systems as passageways to the underworld, and that their wall decorations had either some ritualistic significance, such as to sanctify the space, or were put in place to identify where important resources could be found.

ANCIENT GREEK PAINTING

Despite their fondness for sculpture, the ancient Greeks regarded painting as the highest of the art forms. The main painting surface was the wooden panel. These deteriorated over time and unfortunately no examples have survived.

The first painter to indicate perspective in his works was Polygnotus of Thásos (fifth century BC). His paintings were still being admired six hundred years after his death, but none of his works has survived.

If you could say it in words, there would be no need to paint.
—EDWARD HOPPER (1882-1967)

OIL PAINTING

From the time of the ancient Greeks the chemistries of art and medicine were closely related and were often discussed in the same books.

Oil paint developed from "drying oils," which were originally used for dressing wounds. The oils were also used to form a protective cover or varnish on paintings. Experimentation with pigments enabled color to be added.

The Van Eyck brothers (Hubert, circa 1370–1426; Jan, circa 1390–1441), who came from Maaseyck near Liège in Flanders (present-day Belgium), are credited by some authorities with refining the process, until oil paint provided a flexible medium suitable for painting whole pictures.

CHINESE PAINTING

The Chinese method of drawing or painting, with an unbroken tradition of more than two thousand years, is to use ink or watercolor on silk or paper.

The earliest known example of Chinese painting was excavated from a tomb dating from the time of the Western Han dynasty, which ruled China from 206 BC to AD 9. The painting is known as the Mawangdui banner and is presently in the Hunan Provincial Museum.

Oil painting was introduced to China during the early Qing dynasty (1644–1911) by Jesuit priests employed in the court of the emperor, but it never became more than a novelty.

FRESCOES

The term *fresco* refers to the application of paint to fresh plaster or mortar. As the plaster dries, it absorbs the pigment, which then becomes part of the plaster. As a result, frescoes are far more durable than ordinary paintings.

The earliest frescoes discovered so far date from 1700 to 1400 BC. They were found in King Minos's palace on the island of Crete.

The ceiling of the Sistine Chapel, painted by Michelangelo (1475–1546), is an example of Italian Renaissance fresco work.

MOSAIC

The first known mosaics were terra-cotta cones that were embedded into the outer walls of buildings during the third millennium BC in Uruk (present-day Warka in Iraq). The cones were placed with the blunt end outermost, providing additional protection for the sun-dried bricks, which were the main building material. The red, white, and black cones were arranged to form geometric patterns.

Pebble mosaic was developed during the eighth century BC around

Ankara (in present-day Turkey). It was laid on the floor to provide a hard-wearing surface.

Greek mosaic The ancient Greeks refined the use of mosaic, and during the fifth century BC began to produce floor mosaics as we would recognize them today. The best-preserved examples are in Motya and Morgantina in Sicily.

Roman mosaic began in the second century BC. The Romans largely copied the Greeks and began the process of turning mosaic from an exclusively upper-class art to a commonplace floor decoration.

SCULPTURE

The earliest known of all sculptures or carvings is the Venus of Willendorf, found close to the town of Willendorf in Austria in 1908. Dated at between 10,000 and 25,000 years old, the figure representing a Stone Age woman is carved from limestone, stands only a few inches high, and is understood to be a fertility symbol. The face is obscured but the buttocks, breasts, and genitalia are exaggerated out of all proportion.

There are several other carvings of the same period, Venuses of Kostienski, Maina, Malta, Avdeevo, and others, all of which exhibit the same exaggerated features.

Ivory carvings were produced in ancient Egypt between 4000 and 3200 BC. Civilization in ancient Egypt was highly religious and this was reflected in the subject matter of their carvings, which generally depicted gods and goddesses.

Gigantic monolithic (massive stone) sculptures with a mainly ritual, rather than aesthetic, significance began to appear in ancient Egypt between 3200 and 2780 BC. The Great Sphinx of Giza has been variously dated between 3000 and 2500 BC.

GREEK SCULPTURE
By the fifth century BC the ancient Greeks were creating sophisticated sculptures of the human form. The statues were shown in sporting and

heroic poses, and the leading Greek sculptors were the first to display emotions in their figures. Before the Greeks, the faces of sculptures appeared frozen, without displaying any feeling.

For the first time the names of sculptors became known. Phidias (circa 490–430 BC), who designed many of the works surrounding the Parthenon in Athens, is generally regarded as the greatest Greek sculptor of the classical period.

Praxiteles, probably the greatest Greek sculptor of the fourth century BC, introduced the concepts of grace and sensuous charm into his figures. His statue of the naked goddess known as the Aphrodite of Cnidus was a bold innovation, and was thought by the Roman historian Pliny the Elder (AD 23–79) to be the greatest statue in the world.

ROMAN SCULPTURE

Around 1000 BC the Villanovan civilization, in the region of present-day Bologna, began to produce small bronze and terra-cotta statues for symbolic purposes. Regular changes in style occurred until fine Roman sculptures in bronze and clay, such as the Reclining Couple presently in the National Museum of Rome, emerged in the sixth century BC.

The Romans closely followed the Greeks in their sculpture of the human form. These statues, usually in stern heroic poses, gave the people an opportunity to see what their leaders and heroes looked like, since there was little opportunity to see them in the flesh.

CHINESE SCULPTURE

In ancient Chinese culture any creative activity that involved physical labor was not considered one of the fine arts; sculptors were regarded as mere craftsmen, and few of their names are known. The earliest carved jade figures date from 3400 BC, and there are stone figurines from the eighteenth century BC, the Shang dynasty. They were intended for use as funerary objects, and generally the subject matters were small animals and birds.

China's most famous statues—the serried ranks of eight thousand life-size soldiers known as the Terra-cotta Army—were begun in 240 BC. The first Chinese emperor, Qin Shi Huangdi (259–210 BC), commissioned the sculptures. He died at the age of fifty and the statues of the soldiers in battle-ready formation, all with individually molded facial features and hairstyles, were buried with him to protect him in the afterlife. Work on the statues by 700,000 conscripts took thirty years to complete. Also known as the Terra-cotta Warriors, the statues were chanced upon in 1974 near to the city of Xi'an in central China, by villagers digging in a field to sink a well.

MUSIC

Music is the shorthand of emotion.
—LEO TOLSTOY (1828–1910)

MUSICAL NOTATION

The earliest song to have been written down is a Syrian cult hymn, the Hymn to Creation, written in cuneiform (see "Communication," p. 20) and has been dated at between 3,400 and 4,000 years old.

The ancient Greeks were the first to develop a system of symbols to make a record of musical sounds. Music was annotated by the use of two different systems of letters for instrumental and vocal music. Boethius (AD 470–525) wrote five textbooks on music theory, developing a system of annotation using the first fifteen letters of the alphabet.

Gregorian chant was developed by Pope Gregory I, "the Great" (AD 540–604), during his papacy from AD 590 to 604. The original annotation of Gregorian chant used neumes, which are small marks above the text to indicate the notes in a piece of music. Neumes are thought to have derived from symbols in the Greek language, and modern musical notation is derived from them.

The first published musical score was written in 1581 by Vincenzo Galilei, the father of Galileo, who was a famous performer on the lute and a singer-songwriter.

Scales were invented in the eleventh century by Guido d'Arezzo (circa 991–1033). His system of naming scale degrees used the initial syllables of the lines of a Latin hymn (ut, re, mi, fa, sol, la).

This was the origin of the eight-note scale (the octave) now used in Western music (the tonic sol–fa) and made world famous by the film *The Sound of Music*: do, re, mi, fa, sol, la, ti, do.

ARABIAN MUSIC

Nothing is known of Arabian music before AD 622, but it flourished under various caliphs (Islamic rulers claiming descent from the Prophet, Muhammad) from 661 to 750, using complex tonal colors and constantly changing rhythms.

CHINESE MUSIC

Chinese music dates back to the dawn of civilization in China, and by 1100 BC there was a well-established culture of stylized musical theater, in which music was only one of the elements, and not the principal one. The Chinese Imperial Music Bureau was established between 221 and 207 BC, during the Qin dynasty.

MUSICAL INSTRUMENTS

The first known musical instrument is the flute played by Neanderthal man. The earliest examples have been dated at between 43,000 and 82,000 years old, and one ancient bone flute has been discovered with holes spaced for half tones. The Egyptians, Etruscans, and ancient Greeks all played flutes, sometimes with the nose.

The earliest known drums were excavated in Moravia (eastern part of the Czech Republic) and date from 6000 BC. These examples are made from hollowed-out tree trunks with a membrane of fish or reptile skin stretched across the open ends.

The earliest Chinese musical instrument is a globular clay ocarina (a simple wind instrument) from 5000 BC.

Pan pipes that date from 2500 BC have been found on the Cyclades in the Greek islands.

The first string instrument was the lyre, which dates from 2000 BC in Mesopotamia and ancient Greece. Homer describes Achilles making and playing a lyre, which was used to accompany popular songs in the way the modern guitar is used.

According to the myth, Nero was "fiddling" while Rome burned. In fact he was plucking a lyre, since there were no violins (fiddles) in those days. All string instruments were struck or plucked, until the bow was developed in the Middle Ages.

The organ was invented in 246 BC. We actually know the inventor's name: Ctesibius of Alexandria (circa 285–222 BC). The organ, known as a hydraulis, used the pressure of water to maintain a continuous sound. An example of Ctesibius's organ dating from 228 BC has been discovered close to Budapest, having survived a recent house fire there. The bellows had been destroyed in the fire, but the rest was in good condition. Ctesibius also invented the keyboard, the bellows, compressed air, moving statues, automatic doors, and the clepsydra (a water clock, which was not surpassed in accuracy for more than a thousand years).

The piano was invented in Italy in 1700 by Bartolomeo Cristofori (1655–1732).

The origin of the violin cannot be established with any certainty. The closest dating is the first half of the sixteenth century, when Andrea Amati (circa 1511–80) of Cremona, Italy, was asked to produce a stringed instrument for the Medici family. He was also asked to design an instrument that could be played by street musicians, and came up with the violin. It became an instant success, and some Amati violins from 1564 survive today.

The saxophone was invented in 1846 by Adolphe Sax (1814–94), a Belgian musical instrument maker.

The first electrical musical instrument was the telharmonium (also known as the dynamophone), which was patented in 1897 by lawyer Thaddeus Cahill (1867–1934) of the United States. The telharmonium weighed in at over two hundred tons, and Cahill demonstrated it for the first time in public at Holyoke, Massachusetts, in 1906. He went on to build two further telharmoniums.

The music was created electromechanically, not electronically.

PERFORMANCE

Singers Primitive man sang to invoke his gods and to celebrate rites of passage.

Orchestras Large groups of instruments making up orchestras were a seventeenth-century development. Until that time, music had been played in small ensembles with no need to specify what instrument was to play which notes. As the numbers of musicians grew, orchestration became vital to prevent a complete shambles of sound.

The first example of large-scale orchestration was employed for performances in 1615 of Giovanni Gabrieli's (1555–1612) *Sacrae Symphoniae*, which he composed in 1597. Orchestras as we would recognize them today became fully developed in the late eighteenth century.

Early orchestras that survive today:

Orchestra	Date of Inception
Mannheim	1741
Leipzig Gewandhaus	1742
London Philharmonic	1813
Paris Conservatoire	1828
Vienna Philharmonic	1842
New York Philharmonic	1842
St. Louis Philharmonic	1880
Boston Philharmonic	1881
Berlin Philharmonic	1882
Chicago Philharmonic	1891

Conductors Conducting developed in the Middle Ages and was known as cheironomy, which was the use of hand gestures to indicate melodic shape.

The first well-known conductor was Johann Stamitz of the Mannheim Orchestra in the eighteenth century.

Conductors began to use batons to beat time in the early nineteenth century. Before that, they used a variety of devices including rolled-up papers or a conducting staff, which they would beat rhythmically on the floor. In 1687, Jean-Baptiste Lully (1632–87) of the Paris Opera hit himself in the foot with his conducting staff and died when the wound turned gangrenous.

OPERA

Developing out of Italian music, opera has a long history dating back to Saint Ambrose (AD 340–97), the bishop of Milan and adviser to the Emperor Gratian (AD 359–83). Ambrose imported Syrian musical practices and wrote hymns. He also developed plainsong, now known as Ambrosian chant, out of the Coptic, Byzantine, Jewish, and Hindu chanting traditions.

The Play of Daniel (also known as *Ludus Danieli*) was a twelfth-century musical drama by an unknown composer and is very close to being an opera. It originated at Beauvais Cathedral in northern France and presented familiar episodes from the book of Daniel.

The oldest surviving opera is *Euridice* by Jacopo Peri (1561–1633) and Ottavio Rinuccini (1562–1621). It was performed in Florence in 1600 and had developed from carnival songs and madrigals, which emerged in the fifteenth and sixteenth centuries. Peri and Rinuccini had previously collaborated with Jacopo Corsi (1561–1602) to produce an earlier opera, *Dafne*, but this has not survived.

In the same year another opera, *Rappresentatione di Anima, et di Corpo*, by Emilio de' Cavalieri was performed in Rome.

The world's first opera house built for the purpose was the Teatro Farnese in Parma, Italy. It opened to the public in 1628 after taking ten years to be built.

The first known opera written by an American, was *The Fashionable Lady, or Harlequin's Opera*, which was composed by James Ralph and premiered in London in 1730, but was not performed in the United States. (Note: it was common practice to give plays and operas alternative titles.)

It was another twenty-seven years before an opera written by an American was performed in the United States. The opera in question was *Alfred*, which was composed by William Smith and premiered in 1757.

Ten years later came the first example of censorship of an American opera. In 1767 the world premiere of *The Disappointment, or The Force of Credulity* by Andrew Barton (a pseudonym) was canceled because it allegedly mocked some local Philadelphia dignitaries.

In 1794, Ann Julia Hatton became the first female American librettist when she wrote the words for *Tammany, or The Indian Chief.* In the same year, the first American opera composed by a woman premiered in Philadelphia. The show, called *The Slaves of Algiers, or A Struggle for Freedom* was composed by Alexandra Reinagle, a women's rights campaigner.

Opera's first African American composer was Harry Lawrence Freeman, whose show, *The Martyr*, was first performed in Denver in 1893.

The New York Metropolitan Opera performed its first opera in 1907, *The Pipe of Desire* by Frederick Shepherd Converse.

The first African American artistic director of a major opera company was Willie Anthony Waters, who took control of the sixth oldest opera company in the United States, the Connecticut Opera, on July 1, 1999.

The D'Oyly Carte Opera Company was set up to manage the first Gilbert and Sullivan operettas in the 1880s. In 1887, Richard D'Oyly Carte built the Royal English Opera House, which is now the Palace Theatre, London.

POETRY AND LITERATURE

If poetry comes not as naturally as leaves to a tree,
it had better not come at all.
—JOHN KEATS (1795–1821)

The five main influences from the ancient civilizations were Babylon, Egypt, Greece, Rome, and the Israelites. The Babylonians produced the oldest written narrative, the epic *Gilgamesh*, which dates from 2000 BC and is recorded in verse. The earliest known books are Egyptian papyrus rolls, recording beliefs about the supernatural world, and clay tablets from Mesopotamia.

The earliest poetry in the form of epic poems, such as *Gilgamesh*, seems to have developed out of oral history and storytelling, which began long before man could write. Storytellers would rely on stock phrases, fixed rhythms, and rhyme as aids to memory.

JEWISH (HEBREW) LITERATURE

With a history going back more than three thousand years, the earliest texts in Jewish literature date from 1200 BC. Twenty of the Bible's Old Testament books were written between 1200 and 587 BC. The literature was mostly written in Hebrew, although Greek, Aramaic, and Arabian languages were also used.

The earliest examples of Jewish literature are some of the books of the Old Testament in the Bible and the Apocrypha, the hidden stories, which were handed down as teachings for future generations. The main influence of ancient Hebrew literature came from the writings in the Old Testament of the Christian Bible. Some of the conversations in the Old Testament seem to be attempts to reproduce in writing the style of everyday speech.

The Talmud, one of Judaism's sacred books, was compiled between the first and sixth centuries as the written record of the oral traditions. The Talmud remains the principal authority on Jewish ethics, law, and customs.

GREEK LITERATURE

Modern Western literature derives from the Greek model. Although only a small amount survives, few doubt that the Greeks invented the literary genres of the epic, drama, history (as opposed to chronicling), poetry, and oratory. The first prose writer was Pherecydes of Syros, whose work dates from circa 550 BC.

Greek literature had few influences from other sources.

The oldest of all surviving poems are works in ancient Greek by Homer and Hesiod (circa 700 BC). The *Iliad* and *Odyssey*, attributed to Homer, were probably written down in the mid-eighth century BC. They were the principal Greek records of the Trojan War and arose out of a long oral storytelling tradition.

There is not a great deal known about Homer, although he is generally thought to have been a blind poet who made his home on the Greek island of Chios and may well have dictated the poems to others to write down. Around the same period, Hesiod wrote two epics: the *Theogony*, designed to instruct readers in the ways of the gods, and the *Works and Days*, describing peasant life.

The first great writer of tragedy was Sophocles (496–05 BC), who won the dramatic competition in Athens twenty-four times. He wrote 130 plays of which only seven survive, the greatest of which is considered to be *Oedipus Tyrannus*.

The first great woman poet was Sappho (610–580 BC), who was born on the island of Lesbos. In her work, she expressed feelings of tenderness and passionate love for other women, from which the term *lesbian* has been derived. Historians speculate that Sappho may have written her work not for herself but to help others, particularly less articulate men, who may have been seeking ways to express themselves poetically to admired women (see "Sex," p. 215).

LATIN (ROMAN) LITERATURE

Lucius Livius Andronicus (284–04 BC), the first great Latin writer, was actually a Greek. He translated Homer's *Odyssey* into Latin and, inspired by Homer's massive vocabulary, wrote many other works of his own coining new Latin words and phrases. He was an immense influence on writers who followed, such as Plautus (254–184 BC), who wrote comic plays, and Naevius (264–195 BC), who was an influential dramatic poet.

CHINESE LITERATURE

Chinese literature began around three thousand years ago and has the longest continuous history of any literature in the world. The earliest known texts are records of divinations of the future, carried out for imperial rulers. In contrast to other great literary cultures, the Chinese did not write mythology or great epics. The main subject matter was religious and philosophical thought.

The first anthology of Chinese poetry is the *Shih Ching* containing religious and folk songs. It was produced between circa 550 BC and 480 BC, during the Chou dynasty (1111–255 BC).

The five classics of Chinese literature are: *Shih Ching* (poetry), *I Ching* (changes, fortune telling), *Shu Ching* (history), *Li Ching* (rites), and *Ch'un-Ch'iu* (spring and autumn annals).

JAPANESE LITERATURE

Chinese writing heavily influenced Japanese literature, and, in fact, the earliest Japanese texts were written in Chinese. The first examples date from AD 440 and are inscribed on ceremonial swords.

KOREAN LITERATURE

The earliest known examples of Korean literature are religious songs that were performed before 57 BC. The golden age of Korean literature dates from 57 BC to AD 668, during the period of the Three Kingdoms.

ARABIAN LITERATURE

The earliest known form of Arabian literature is the heroic poetry of the so-called noble tribes of pre-Islamic Arabia (before AD 622). This time is known to Muslims as the Jahiliyyah or the Period of Ignorance, and the literature was recorded only two centuries later, in the *Mu'allaqat*, a group of seven long poems, and the *Mufaddaliyat*, a collection of 126 poems dating from AD 500.

Often considered as the greatest of the poets of the Jahiliyyah was Zuhayr (AD 520–609), who wrote in minute detail about the everyday life of the Bedouin. The standard Arabic verse is the *qasidah*, a long poem, often reciting incidents from the poet's life or that of his tribe.

Pre-Islamic poetry was preserved orally until the late seventh century when Arab scholars undertook the mammoth task of collecting and recording verses that had survived only in the memories of the reciters. The golden age of Arabian literature began after the rise of Islam in 622.

ENGLISH LITERATURE

The origins of English literature are found in the Old English alliterative verse of Caedmon, who wrote the Hymn of Creation in the seventh century. Caedmon was mentioned in *The Ecclesiastical History of the English People* written by the Venerable Bede (AD 673–735) in the eighth century. According to Bede, Caedmon was an uneducated herdsman who received a divine call in later life, became a monk, and began to write poetry in vernacular language.

The oldest surviving Anglo-Saxon epic poem is *Beowulf*. It was written sometime around the tenth century and describes the adventures of a Scandinavian warrior of the sixth century. *Beowulf* is the earliest poem in what could be described as early English, and the only surviving text was almost destroyed when the Cotton library at Ashburnham House caught fire on October 23, 1731.

The first novel written in English is said to be *The Unfortunate Traveller, or The Life of Jacke Wilton*, a picaresque story written by Thomas Nashe (1567–1601) and published in 1594.

The first detective novel is widely regarded to be *The Woman in White*, written in 1860 by Wilkie Collins (1824–89). Collins was famous for his formula for writing a bestselling novel—not giving the end away until the very last moment. The principal reason for this was Victorian publishers' requirement that novels be published in three volumes. No hint of a resolution could be given until the third volume had been produced. Collins summed up the principle when he said: "You make 'em laugh, you make 'em cry, but most of all . . . you make 'em wait."

IRISH LITERATURE

The earliest surviving examples of Irish literature are the ogham inscriptions of AD 300–500. These inscriptions were carved into tombstones and other funerary items in Celtic, which had been introduced to Ireland in the third century BC.

Ogham script is a unique writing system that uses sets of one to five dots for vowels and combinations of parallel lines for consonants. The origin of ogham is uncertain, but according to Irish legend it was created by the Irish god Ogma.

The earliest known Irish poem is a eulogy to Saint Columba (AD 512–97) thought to have been written by Dallan Forgaill, the chief poet of Ireland, in the eighth century.

The famous illustrated manuscript known as the Book of Kells was begun around AD 750 in the monastery of Iona. The manuscript contains the four gospels and lists of Hebrew names. It is thought that the book was most likely captured by Vikings and transferred to Kells in County Meath and completed in the ninth century.

> *Poets aren't very useful.*
> *Because they aren't very produceful.*
> —OGDEN NASH (1902–71)

AMERICAN LITERATURE

The relatively recent "discovery" and settlement of the continent of America by Europeans and others is the reason American literature was late on the world stage. The first known important letter written from either North or South America was sent by Hernán Cortés (1485–1547) to the Spanish Crown in 1519. The letter, which was the first of five *cartas de relación* written by Cortés, has since been lost, and his second letter, of 1520, is the first known survivor. The letters ran to many pages and provided a detailed account of the conquest of the Aztecs, and the settlement of Mexico.

William Bradford (1590–1657) crossed the Atlantic on the *Mayflower* in 1620, as one of the first generation of pilgrims to leave Europe for the New World. Bradford had been one of the organizers of the expedition, and en route to the New World, he helped to draft the so-called Mayflower Compact, which was intended to provide a code of living and ethics for the new settlers.

The Mayflower Compact begins: "We whose names are underwritten, the loyal subjects of our dread Sovereign Lord King James, by the Grace of God of Great Britain, France, and Ireland King, Defender of the faith, etc." Within a few months of the colony's establishment, Bradford was elected governor. Bradford gives a vivid account of the early years of the settlement, and the daily challenges faced by the settlers, in his *History of Plymouth Plantation*, which he wrote between 1620 and 1647. Although the book was not published until more than two hundred years later, in 1856, Bradford, who also wrote long descriptive poetry, has a rightful claim to be the first figure in American literature.

Anne Bradstreet (1612–72), who was married to Simon Bradstreet (1604–1697), the governor of Massachusetts, is regarded as America's first English language poet. Bradstreet's first volume of poetry, *The Tenth Muse Lately Sprung Up in America,* was published in London in 1650 by her brother-in-law, without her permission.

In 1653 John Eliot (1604–90) published *Catechism in the Massachusetts Indian Language,* the first book ever published in a Native American language. Between 1661 and 1663, Eliot translated the Bible into the Massachusetts Indian language, which became the first Bible printed in North America.

DANCE

*Dance is the only art of which we ourselves are
the stuff of which it is made.*
—TED SHAWN (1891–1972)

The earliest dance is shown in the cave paintings that date from 40,000 to 10,000 BC at Les Trois Frères in southern France. Half-human figures wearing animal costumes are shown in dancing poses.

From as early as 5000 BC there are clay figurines shown with their hands raised above their heads, which indicate that dance formed part of religious activity in ancient Egypt. Tomb carvings of around 3500 BC depict masked dancers with the priest or king dancing to represent a god.

The oldest surviving European dance is the Austrian-Bavarian Schuhplattler (the shoe-slapping dance), which is thought by historians to date from Neolithic times, around 3000 BC. In the Schuhplattler, the man lifts his feet to knee height to slap his shoes, and the woman spins around on the spot.

BALLET

Although the Romans did have a sort of ballet called the *Fabulae Atel-lanae*, it was not until the sixteenth century that ballet began to emerge out of lavish masquerade balls where everyone wore a mask and mum-meries, festivals during which people went about the streets in disguise.

The first ballet combining all the elements of movement, music, decor, and special effects was *Le Ballet Comique de La Reine*, which was performed in 1581 for Catherine de Médici (1519–87), Queen of France. The ballet was performed at the Valois court and is the first known work to have combined dance, verse, and music into a coherent whole.

The first set of principles governing ballet was *Orchesographie*, writ-ten by Thoinot Arbeau in 1588. The first ballet school was the Academy of the Art of Dancing, started in Paris in 1661 by Louis XIV, King of France.

The first time women were allowed to dance ballet in public was in 1681 at the Paris Opera (strangely, this is the name of the Paris ballet company).

The artistic positions of ballet were set out in *Letters on Dancing and Ballet*, published in 1760 by Jean-Georges Noverre (1727–1810). Noverre was the major modern reformer of ballet, and by 1773 it had developed into the form of dance spectacle performed today.

COMMUNICATION

COVERING PERSON-TO-PERSON
AND BROADCAST COMMUNICATION

Since messages were first left on tablets of stone, man has used every means at his disposal to communicate private, public, and secret messages to others. Immense fortunes have been made, and countries conquered, by the controllers of communication.

PERSON-TO-PERSON

WRITING

As with all ancient technologies, it is impossible to date the origins of writing precisely. Leaving scratches on rocks with the use of flint hand tools was almost certainly primitive man's first attempt at a form of coded communication.

The first written language was cuneiform, which was developed by the Sumerians of southern Mesopotamia more than five thousand years ago (see "Language," p. 164).

The earliest hieroglyphs were discovered carved into burial stones in Egypt and date from 3200 to 2950 BC (the predynastic period). In 1894, two English archaeologists, James Quibell (1867–1935) and Frederick Green (1869–1949), working at Nekhen in Upper Egypt, discovered the Narmer

Palette, a type of shield. For many years the Narmer Palette was the earliest evidence of hieroglyphic writing, but recent archaeological discoveries have revealed hieroglyphic symbols on earlier Gerzean pottery that have been dated provisionally at circa 4000 BC. The Gerzean people were prehistoric dwellers along the west bank of the Nile.

Until 1799 hieroglyphs were impossible to translate, but in that year Napoleonic soldiers stationed close to the town of Rosetta discovered a large black basalt stone. Frenchman Jean-François Champollion (1790–1832) examined the stone and realized that it bore the same inscriptions written in Greek, hieroglyphs, and demotic (the language of the common Egyptian people). Since Greek was a known language, it then became possible to translate the hieroglyphs.

The earliest known alphabet was devised in Ugarit, in present-day Syria, around 1500 BC. Discoveries at Ras Shamra in 1929 unearthed the Ugarit alphabet.

The earliest Chinese writing dates from 1500 BC and is inscribed on pieces of bone and tortoiseshell. The system of writing the Chinese language has changed little since then, as it still relies on pictograms and characters.

Pens

The first pens were developed around 1000 BC from the brushes normally used by the Chinese for writing.

In 300 BC the Egyptians used thick reeds as pens. The reed had to be continuously dipped into the writing material (a form of ink), which dried quickly in use.

A bronze pen was found in the ruins of Pompeii.

Quill pens were used in the seventh century by Saint Isidore of Seville (circa AD 560–636), although feathers were probably used at an earlier unknown date.

The first machine-made pen nib was produced by John Mitchell of Birmingham, England, in 1828.

The earliest known example of a fountain pen was produced by a M. Bion of Paris in 1702.

The fountain pen was patented in the United States in 1884 by

Lewis E. Waterman (1837–1901), although he did not actually invent it. Robert Thomson (1822–73), the inventor of the pneumatic tire, obtained a British patent for the principle of the fountain pen in 1849.

The ballpoint pen was patented by John J. Loud on October 30, 1888, but he failed to exploit his invention, using it only for marking leather. Loud's patent was ignored in later battles over rights to claim the invention.

The first successful mass-produced ballpoint pen was developed and patented in the 1930s by Lazlo Biro (1899–1985), a Hungarian living in Argentina, who also invented the automatic gearbox for cars. The pen became universally known as the Biro.

Baron Marcel Bich introduced another mass-produced ballpoint pen, the Bic, in 1950.

Soft-tip pens using porous material for the nib were introduced in the 1960s.

Ink

The Chinese developed a type of ink around 3000 BC. It was made from a mixture of soot from pine smoke, lamp oil, and gelatin from animal skins, and was applied by brush.

Pencils

The modern lead pencil only became possible after the discovery of a deposit of pure graphite in Borrowdale, England, in 1564, and was first described by Swiss naturalist Conrad Gesner in 1565. Early versions were made by wrapping graphite in string, but by 1662, wood pencils were being mass-produced in Nuremberg, Germany. William Monroe of Concord, Massachusetts, made the first American wood pencils in 1812.

The lead pencil is misnamed, since it contains no lead. The name originated from lead wheels that were used to produce straight lines on paper or parchment. The marks looked remarkably like those from the pencil and hence the pencil became known as the "lead" pencil.

The pencil eraser, also known as a rubber, was invented by an English engineer, Edward Naime, in 1770.

The pencil sharpener was invented by Therry des Estwaux of France in 1847.

Paper and parchment

Papyrus The ancient Egyptians first used a form of writing material called papyrus from about 2400 BC. Although the word *paper* is derived from *papyrus*, the two products are fundamentally different. Papyrus is made from sheets of thinly cut strips from the stalks of the papyrus plant (*Cyperus papyrus*), whereas paper is made of the matted fibers of several different plants, which are soaked, macerated, and formed into sheets.

Shortages of papyrus led to the development of parchment, which derives its name from the ancient Greek city of Pergamum (Bergama in present-day Turkey). It is supposed to have been invented in the second century BC, and was produced from the skin of specially bred cattle.

The earliest paper is said to have been invented by Ts'ai Lun in China in AD 105, but earlier papers have been discovered. By AD 750, paper was being used in Samarqand, and by AD 794 it had spread to Baghdad and then on to the rest of the world.

The first papermaking machine was produced by Nicholas-Louis Robert of France in 1798.

WRITING SYSTEMS

Around 600 BC there was general consensus among Mediterranean cultures to adopt left-to-right, top-to-bottom writing and reading. Before that, there had been a mixture of right to left, bottom to top, and even boustrophedonic, which involved writing in opposite directions on alternate lines.

Shorthand

The earliest known shorthand was used during the fourth century BC by the ancient Greeks. They used a system of symbols in which a single stroke could represent complete words. The word *stenography* is derived from the Greek words, meaning "narrow writing," used to refer to this system.

In Rome, in 63 BC, Cicero's secretary, Tiro, devised a system of simplified letters and symbols to help him record Cicero's speeches in the earliest organized system of shorthand. It was known as *notae Tironianae* and was still being used in Europe into the Middle Ages.

The first modern shorthand was devised by Sir Isaac Pitman (1813–97). Called Pitman shorthand, it was published in 1837. Before the use of dictation machines, this system enabled notes to be taken at talking speed. Pitman's brother, Benn, who had settled in Cincinnati, introduced the Pitman system to the United States.

The speed record for taking shorthand was set in 1922 when Nathan Behrin, competing in the annual National Court Reporters Association Speed Contest in New York, succeeded in taking 350 words per minute for a sustained two-minute period. The record still stands.

Printing

The first printing process was developed by the Chinese by the second century AD. It was a simple form of fixed-type printing in which the type was permanently fixed to the printing head. Between 1041 and 1048 the Chinese alchemist Pi Sheng developed movable type in which the letters were changeable, but the complexity of the language discouraged further development.

The modern printing press as we know it today was invented by Johannes Gensfleisch Gutenberg (circa 1400-68) in Germany.

The first printed book in English was *The Recuyell of the Historyes of Troye* printed by William Caxton (1422-91) on a printing press in Bruges, Belgium, in 1474-75.

The first book known to have been printed in England was the *Dictes or Sayengis of the Philosophres* printed by William Caxton after he had set up a printing press in Westminster, London, in 1476.

MAIL DELIVERY

Pigeon post was used first by the Sumerians in 776 BC.

The first royal mail in the United Kingdom was introduced by Henry VIII (1491-1547) in 1516 when he appointed Sir Brian Tuke as his master of the posts. In theory the service was for everyone, but the general public was discouraged from using what was effectively the king's private mail service. The Royal Mail as we know it today was established in 1635, and the Royal Charter was granted on September 26, 1839. Postcoding of all addresses in the United Kingdom was completed in 1974.

The earliest reference to a continuous message relay system that employed a horse and rider is from ancient Egypt around 2000 BC. Cyrus the Great, in sixth-century BC Persia, had permanent post houses set up at intervals along mail routes to service both horses and riders.

The Pony Express began operating on April 3, 1860, between Missouri and Sacramento, California, and ran its last mail in 1861, shortly after the overland telegraph had been completed in October of that year. During its short, romantic existence, the Pony Express was the fastest means of delivering messages across the United States.

Airmail

The world's first airmail service began in 1911 when "aerial post" was flown from Hendon to Windsor in England.

The first overseas airmail was established in 1919 between London and Paris.

The first transatlantic airmail link was established only six weeks after Charles Lindbergh made his historic first solo crossing of the Atlantic. On June 29, 1927, U.S. airmen Richard E. Byrd, Bert Acosta, Bernt Balchen, and George Noville took off from Roosevelt Field, New York, to fly to France in a Fokker C-2. Thick fog in Paris forced them to change course and they had to ditch in the sea three hundred yards off the beach at Ver-sur-Mer. The mail got damp but was still delivered.

SEMAPHORE

A system of signaling by holding flags or lights in a particular pattern, semaphore was originally developed in 1794 by the Frenchman Claude Chappe (1763–1805) to transmit messages quickly across long distances on land or sea. On land, observers were sited five to ten miles apart to receive and send messages, and the system became the standard form of ship-to-ship communication before radio.

TELEGRAPH AND TELEGRAMS

The first electromagnetic telegraph was invented in 1837 by two English physicists, Sir Charles Wheatstone (1802–75) and William Cooke (1806–79).

Morse code was invented in 1838 (together with the Morse telegraph) by Samuel Finley Breese Morse (1791–1872). On January 6, 1838, Morse succeeded in sending the first private telegraphic message along a three-mile-long wire stretched around a room. The message was, "A patient waiter is no loser."

International Morse code (a modified, simpler version) was developed in 1851.

The first public telegraph line was erected in 1843 between Washington DC and Baltimore, a distance of forty miles. Samuel Morse sent the first message on May 24, 1844, over an experimental line. It said, "What hath God wrought?"

The first telegram was transmitted on April 8, 1851, when a group of businessmen formed the New York and Mississippi Valley Printing Telegraph Company. It started immediately with 550 miles of wire and a license to use a printer invented by Royal E. House. The device was the first to print letters and numbers instead of the dots and dashes of Morse code, thus enabling the transmission of the first telegram.

Later, the New York and Mississippi Valley Printing Telegraph Company changed its name to Western Union.

The first undersea telegraph cable was laid between France and England in 1850. The cable was cut by the anchor of a French fishing boat after three days, and was not replaced until the following year, when a more substantial cable was laid. A previous mile-long experimental cable had been laid between HMS *Blake* and HMS *Pique* under the waters of Portsmouth harbor.

The first transatlantic cable was laid in 1858 by Massachusetts-born Cyrus Field (1819–92), but it only lasted a month. The cable ran between Newfoundland and Ireland, and Queen Victoria sent President Buchanan a telegram of congratulation on its opening on August 16, 1858.

TELEX

Connected by telephone lines, telex is a system of linked teleprinters (electromechanical typewriters), which allows trained operators to transmit and receive typed words simultaneously. Twenty-five messages could be transmitted simultaneously on a single telephone connection, which for a long time made telex the cheapest form of long-distance communication.

The global telex network was put in place in the 1920s, and although it was mostly superseded by fax in the 1980s more than 3 million telex lines remain in use worldwide.

The predecessor of the telex system was the stock ticker of 1870, which printed share prices and other financial information onto a continuous strip, direct to the offices of stockbrokers, banks, and other financial institutions. It was called the stock ticker because of the continuous ticking sound it made as the information was printed.

FAX

The fax was invented by Scottish philosopher and clockmaker Alexander Bain (1811–77) and patented as long ago as 1843. Using his knowledge of pendulums, Bain produced a line-by-line scanning mechanism. The idea behind the patent was for the message to be scanned and transmitted line by line along telegraph wires.

The first commercial fax machine, the pantelegraph, was sold by Giovanni Caselli in 1861, before the telephone was in common use and messages could actually be carried over public telephone wires. The pantelegraph entered service in 1863, between Paris and Marseilles, and in the first year of operation more than five thousand faxes were transmitted.

E-MAIL

The first use of e-mail was in 1971 on the ARPANET (see also "Internet," p. 34). It was developed by Ray Tomlinson (b. 1941). The first e-mail message was "qwertyuiop" (the letters on the top line of a keyboard), and Tomlinson chose the @ symbol to denote which user was "at" which computer. Asked how he had come to invent such a thing as e-mail when there was no known demand for it, Tomlinson said, "Because it seemed like a neat idea."

TELEPHONE

The telephone was invented in 1876 by Scotsman Alexander Graham Bell (1847–1922) at the age of twenty-nine after he had emigrated to Boston. He had succeeded in transmitting speech sounds the year before.

Elisha Gray (1835–1901) also claimed to have invented the telephone in 1876, but Bell beat him to the patent office by just a few hours. A major court battle followed, which was decided in favor of Bell. Gray had produced a working prototype in 1874 but neglected to patent it.

The first microphone was part of the telephone transmitter, which was invented in 1876 by Emile Berliner (1851–1929) while he was working for Bell.

MOBILE PHONE
The use of a form of mobile telephone (two-way radio) was pioneered by the Chicago Police Department in the 1930s to stay ahead of Prohibition gangsters. The Untouchables, led by Eliot Ness (1903–57), were the first users.

The mobile phone (also known as the cell or cellular phone) as we know it was invented by Dr. Martin Cooper of Motorola. It was first used in 1973 in a demonstration call, which was made by Cooper to his rival, Joel Engel, the head of research at Bell Laboratories.

Hedy Lamarr, the 1930s Hollywood beauty, invented the frequency-switching system that allows cell phones to work. The device works by rapidly switching the signal between the frequency channels, which are recognized by both the transmitter and receiver. Her patent ran out before it was developed commercially, but if she had lived long enough, and not lost the patent rights, she could have been one of the wealthiest women on earth. (See "Sex," p. 214 and "War," p. 260.)

By 1977 Bell Laboratories built a prototype cell system, which was tested by two thousand selected customers. In 1979 the Japanese tested mobile phones in Tokyo.

BROADCAST COMMUNICATION

NEWSPAPERS
The first official publication of news was in the *Acta Diurna* (*Daily Acts*) instituted in 59 BC by Julius Caesar in Rome. They were a record of daily goings-on, such as births, marriages, divorces, and deaths, engraved

on metal or stone and displayed in public places for the general populace to read. They continued to be displayed for the next three hundred years.

A form of *Acta Diurna* had previously appeared in 131 BC, but was merely a record of the outcome of trials and legal proceedings.

The first Chinese newspapers were posted in public places in the eighth century AD.

The *Anglo-Saxon Chronicle* was first published in 890 at the request of Alfred the Great (AD 849–99) to keep a year-by-year record of events rather than publish news. The *Chronicle* actually starts at the year AD 1, and it was maintained until the middle of the twelfth century at ecclesiastical centers such as Peterborough Abbey.

The *Notizie Scritte* was published in 1556 in Venice, which was then a city-state. The public paid a small coin: a *gazetta*. Hence the common newspaper title *gazette*.

The *Gazette de France*, first published in 1631, was the first French newspaper.

The first and oldest magazine still being published anywhere is the *London Gazette*, which was first published in 1665 in Oxford. In its present form the *London Gazette* is a court magazine and no longer publishes general news.

The first British daily newspaper was the *Daily Courant*, which was published from 1702 to 1735.

Le Figaro was first published in 1826 in Paris.

The first American newspaper was *Publick Occurrences, Both Forreign and Domestick*, which was published in 1690 in Boston by Benjamin Harris, an English radical who had been jailed and pilloried (see "Crime," p. 60) in England for publishing seditious pamphlets. Only a single issue was printed before the paper was suppressed by the colonial governor.

Although the *Boston News-letter*, which was published by the postmaster, was launched in 1704, it is the *New England Courant* of 1721, published by Benjamin Franklin's older brother James, which is regarded as the first *independently* published newspaper in the United States.

New York got its first newspaper in 1725 with the publication of the *New-York Gazette*. By the outbreak of the War of Independence in 1775 there were thirty-seven other city newspapers.

The first city newspaper to change from weekly to daily editions was the *Pennsylvania Evening Post*, in 1783, and in 1785 the short-lived *New York Daily Advertiser* became the first to be published daily from its inception.

Lifting of the censorship laws in 1801 allowed newspapers to conduct public campaigns and to criticize the government. In this way, the U.S. newspaper industry pioneered the character of newspapers worldwide.

The *Financial Times* was first published on February 13, 1888, and sold for one penny. It was the first daily financial paper and incorporated an earlier publication, the *London Financial Guide*, which had been published weekly. The businesses covered on the title page of the first *FT* included:

- Welsh gold mining reporting the discovery of 2,500 tons of gold ore
- The American Machine–made Bottle Company raising £600,000
- The Freshwater, Yarmouth and Newport Railway Company (Isle of Wight), which was inviting subscriptions for debentures
- Swan United Electric Light Company reporting annual profits (see "Inventions," p. 156)

The first tabloid newspaper was the *Daily Mirror*, which was published in 1905 by Alfred Harmsworth, later Lord Northcliffe.

The word *tabloid* was coined by medical benefactor Sir Henry Wellcome (1853–1936), and registered as a trademark in 1884. It denotes paper size, not the type or style of content.

The first American tabloid newspaper was the New York *Daily News*, published in 1919. It was, and still is, devoted to sex and sensationalism.

Japan's first daily newspaper, the *Yokohama Mainichi Shimbun*, was published in 1870.

The first Linotype-set newspaper was the *New York Herald Tribune*, in 1886. Until then printing type had to set by hand, but the startling development of the Linotype machine, invented by Ottmar Mergenthaler (1854–99), increased the speed of typesetting by a factor of six.

The *Newcastle Chronicle* was the first British newspaper to install a Linotype machine, in 1889.

The first computer-based printed newspaper was *Today*, published in 1986 by Eddie Shah (b. 1944). It was also the first newspaper to be printed in color. New computer-based print technology developed rapidly in the 1970s and 1980s, enabling journalists to input articles directly into the print room, and editors to move stories about on a computer screen to fit the page. Shah's breakthrough in the use of print technology led to the breaking of the power of the printworkers' union. *Today* ceased publication in 1995.

The first successful daily newspaper comic strip, *Mutt and Jeff*, emerged in 1907 in the *San Francisco Chronicle*. The cartoon was syndicated worldwide and ran uninterrupted until 1982.

The Yellow Kid was included in a comic section in William Randolph Hearst's (1863–1951) *New York Journal American* in 1897. The cartoon lasted only until 1899, having first appeared for a few occasions in *Truth* magazine in 1894. The creator of *The Yellow Kid*, Richard Outcault, pioneered speech balloons.

The *New Statesman* was first published in 1913 by Sidney (1859–1947) and Beatrice Webb (née Potter) (1858–1943).

The *New Yorker* was first published in February 1925 by Harold Ross (1892–1951). Ross edited every edition until his death—a total of 1,399 issues.

The *Sun* was first published in 1963 as the successor to the defunct *News Chronicle*.

The *Independent* was first published in 1986, and is the most recent of the current national daily newspapers to be launched in the United Kingdom.

NEWSREELS

The first newsreels were shown by Frenchman Charles Pathé (1863–1957). He began showing short films of current events (a reel of news) in the music halls of France in 1909. He expanded the idea into Britain and the United States in 1910. Pathé News still exists as Associated British Pathé.

RADIO

Michael Faraday (1791–1867) demonstrated the theoretical possibility of radio communications when he proved that an electric current could produce a magnetic field, and as early as 1864, James Clerk Maxwell (1831–79) proved mathematically that electrical disturbances could be detected over considerable distances.

Radio waves were first detected in 1884 by the German physicist Heinrich Hertz (1857–94), who proved Maxwell's theory over a short distance at Bonn University. Hertz originally named the waves he identified Hertzian waves, but they were soon to be renamed radio waves.

The father of radio is widely regarded to be Guglielmo Marconi (1874–1937). By 1901 Marconi had perfected a radio system, which was tested by transmitting Morse code across the Atlantic.

However, in doing so he infringed seventeen patents previously registered by Serbian-born Nikola Tesla (1856–1943). Marconi's U.S. patent applications were initially turned down because of Tesla's previous work in 1893, but in a major about-face, Marconi was granted a patent for the invention of radio in 1904. Despite the award of the Nobel Prize for Physics to Marconi in 1909, the patent for radio was reversed in favor of Tesla in 1943, a few months after Tesla's death.

Nikola Tesla was born in Serbia in 1856 and granted American citizenship in 1891. He was a prolific inventor who worked with Thomas Edison and George Westinghouse, and held more than seven hundred patents.

The first voice broadcast on radio was a message sent from Brant Rock, Massachusetts, on Christmas Eve in 1906, to shipping in the Atlantic. The speaker, Reginald Fessenden (1866–1932), had previously transmitted from station to station (as opposed to broadcast) across the Potomac River in 1900.

The first live broadcast of opera on radio was in 1920 from Marconi's Chelmsford factory. The Australian soprano Dame Nellie Melba (1861–1931) performed in the opera.

The first transmission from the British Broadcasting Company Limited, which was set up as a commercial company, was on November 14, 1922, from Marconi House in London. By the end of 1922 there were four employees, including John Reith (later Lord Reith 1889–1971), the first managing director. By the end of 1925 there were more than six hundred staff members. In 1926 the name was changed under a noncommercial Crown Charter to the British Broadcasting Corporation (BBC).

The first broadcast of the BBC pips as a time signal, more properly known as the Greenwich Time Signal, began in 1924, and they have been broadcast ever since.

There are five pips of a tenth of a second each and a final pip of half a second long. The hour changes at the start of the final pip. Every few years there is a "leap second," when a seventh pip is added to take account of the slowing down of the earth's rotation.

The first commercial radio station was KDKA in the American city of Pittsburgh, which began broadcasting in 1920.

TELEVISION

In 1884, the Prussian Paul Gottlieb Nipkow (1860–1940) invented a means of transmitting pictures by wire, using rotating metal discs. Nipkow was granted a patent at the Imperial Patent Office in Berlin, but was never able to demonstrate his system.

Television was first demonstrated by John Logie Baird (1888–1946) in 1926 at Selfridges in London, using a mechanical system of rotating discs, which had been successfully patented in 1924. Baird's system was adopted by the BBC in 1929, although the wholly electronic American system, invented by American physicist Philo T. Farnsworth (1906–71), replaced it in 1937.

The first public display of television in the United States was on April 30, 1939, at the opening of the New York World's Fair. The National Broadcasting Corporation (NBC) announced at the fair that it was ready to begin broadcasting television programs for two hours every week. CBS also began broadcasting in 1939, and by the middle of 1940 there were twenty-three TV stations in the United States.

The regulation of television frequencies is controlled by the Federal Communications Commission, which was created in 1934.

The first intercontinental television broadcast was an experimental fifteen-minute telecast on July 22, 1962, from America to Europe using the Telstar satellite (see "Space," p. 225).

The first successful color television system, designed by the Radio Corporation of America (RCA), began broadcasting on December 17, 1953.

The first program broadcast in color was an episode of *Dragnet*, a hard-boiled detective series.

The first coast-to-coast color television broadcast in the United States was on January 1, 1954. It showed the annual Tournament of Roses Parade in Pasadena, California.

MICROPHONES

The first carbon microphone was invented in 1878 by David Edward Hughes (1831–1900), but it was Sir Charles Wheatstone (1802–75) who coined the word *microphone* in 1827.

An earlier version of the microphone was invented by Emile Berliner while working for Alexander Graham Bell (see "Telephone," p. 27).

VIDEO

The first videotape recorder was the Ampex VRX-1000. It was introduced for use in the broadcasting industry in 1956 in Chicago at the National Association of Radio and Television Broadcasters convention.

The first domestic videocassette recorder was developed by JVC of Japan.

THE INTERNET

The forerunner of the Internet was the ARPANET. ARPA stands for Advanced Research Projects Agency, a division of the U.S. Defense Department that possessed linked computers across North America and wanted them to exchange information.

The ARPANET was planned in 1966, started working in 1969, and ceased operations in 1990.

The World Wide Web was invented in 1989 by Tim (now Sir Tim) Berners-Lee and his colleagues at CERN, an international scientific establishment based in Switzerland. They developed HTML (hypertext mark-up language), the format that tells a computer how to display a Web page, HTTP (hypertext transfer protocol) that allows clients and servers to communicate, and URLs (universal resource locators) that identify resources on the Web, such as documents, images, downloadable files, services, and electronic mailboxes, and reduce a range of complex instructions to a single mouse click. Vannevar Bush (1890–1974) of the United States, as director of the Office of Scientific Research and Development, and an adviser to President Truman, had proposed the basics of hypertext as early as 1945.

The World Wide Web provides the facility for all the world's computers to be linked, making it easy to send documentation electronically via the Internet. Tim Berners-Lee deliberately withheld from patenting his work on HTTP, HTML, and URLs because he wanted to encourage as many people as possible to use the Web.

COUNTRIES
AND EMPIRES

COVERING COUNTRIES, OLD EMPIRES,
AND NEW EMPIRES

Men may be linked in friendship.
Nations are linked only by interests.
—ROLF HOCHHUTH (B. 1931)

COUNTRIES

UNITED STATES OF AMERICA

The first inhabitants of what is now North America were Asians who crossed the land bridge that then existed between Asia and Alaska, and spread through Canada and into the United States. There is very little archaeological evidence, but what there is suggests that this migration occurred between 35,000 and 15,000 years ago. DNA analysis links Asians and Native Americans, supporting the view that it was Asians who were the first to migrate to the New World.

Analysis of the three main Native American language groupings, Amerind, Na-Dene, and Eskimo-Aleut, led researchers initially to believe that there were three main migration waves. However, recent language analysis, using a Cray supercomputer, suggests that many more migration waves took place.

It is widely believed that the Genoese Christopher Columbus (1451–1506) was the first European to sail to America, in 1492, although he most likely landed on the Caribbean island of San Salvador (see "Questionable Origins," p. 189).

The first permanent European settlement in America was at St. Augustine, Florida, set up by Don Pedro Menendez (1519–74) on September 8, 1565.

According to legend the whole of America was named after an Italian explorer and mapmaker, Amerigo Vespucci (1454–1512). There is also disputed evidence that America was named after Richard Amerike (1455–1503), a Welshman also known as Richard ap Meryk. Amerike sponsored John Cabot's 1497 voyage, which led to the discovery of Newfoundland. Amerike had specifically requested that any newfound lands were to be named after him. It is thought possible that John Cabot (circa 1425–1500), knowing of Columbus's discovery in 1492, crossed the short distance to the east coast of America from Newfoundland and returned to Britain with the new land named after his sponsor. Interestingly, Amerike's family coat of arms is made up of stars and stripes, almost identical to the U.S. flag (see "Questionable Origins," p. 189).

The first European to sight California on the West Coast of the North American continent was Sir Francis Drake (1540–96); he claimed the land for England in 1579.

AUSTRALIA

The first inhabitants of Australia were the so-called Aborigines who arrived on the continent between 40,000 and 60,000 years ago from the Asian mainland. They most likely arrived either by a now-submerged land bridge or by rafts and canoes.

The first Europeans to visit Australia were the Dutch, in 1606, although there may have been earlier sightings by the Portuguese. In 1688 the English pirate and travel writer William Dampier landed on northwestern Australia as captain of the *Cygnet*.

The first European to claim possession was Captain Cook (1728–79) when he stood on Possession Island on August 15, 1770, and claimed eastern Australia for England and King George III. Cook had made his first landfall in the *Endeavour* on April 29, 1770.

The first European settlement was established in 1788 when the British First Fleet, under the command of Captain Arthur Philip (1738–1814) sailed into Botany Bay with more than a thousand prisoners and officers. Port Jackson was considered more suitable, so they moved down the coast, and Philip became the first governor of what was then named New South Wales.

The first free settlers were Thomas and Jane Rose of Dorset, who landed in Australia on January 15, 1793, with their four children. The Roses' second home, Rose Cottage, is still standing today.

The European settlers gradually occupied the land that had been the ancestral home of the Aborigines. Although they were grouped under the name Aborigine, there were up to seven hundred distinct groupings of native inhabitants, with widely differing societies, languages, and cultures.

NEW ZEALAND

The first inhabitants of New Zealand were the Maori, who, it is estimated, migrated there at some time around AD 800, although they may have arrived much earlier. This makes New Zealand the most recent of the Pacific Islands to be occupied.

The first European to discover New Zealand was the Dutch explorer Abel Tasman (1603–59), in 1642. Earlier that year, Tasman had discovered Tasmania, the island that is now named after him, although he started by naming the island Van Diemen's Land after his sponsor. Tasman then sailed east and on December 13 discovered the west coast of the South Island of New Zealand, which he named Staten Landt. The name was changed to Zelandia Nova (New Zealand) after the Dutch province of Zeelandt. It is thought that a group of directors of the Dutch East India Company made the final naming decision.

Captain Cook (1728–79) charted New Zealand in 1769 on his way back from Tahiti. Cook had been visiting Tahiti to observe the transit of Venus across the sun for the Royal Observatory.

Until 1792 there were two New Zealands. The other New Zealand was an island off the coast of New Guinea (the western half is Indonesian, known as Irian Jaya, and the eastern half is British, now known as Papua New Guinea). The name of the second New Zealand was never formally changed but rather fell into disuse.

EIRE (SOUTHERN IRELAND)

On Easter Monday in April 1916, the Easter Rising took place in Dublin, in which the Irish Volunteers and the Irish Citizen Army seized key locations across the city. The revolt was a protest against British rule and an attempt to declare Ireland a republic.

In the immediate aftermath of the Easter Rising, great public sympathy was generated after the British executed several of the leaders. This led to the Irish War of Independence, which lasted from 1919 to 1921, and to the Anglo-Irish Treaty, which was signed on December 6, 1921. The treaty gave independence to twenty-six of the thirty-two counties of Ireland, which were then grouped under the name of the Irish Free State.

The Irish Free State became Eire after the constitution was formally adopted by plebiscite in 1937.

OLD EMPIRES

The concept of empire refers to an extended territory dominated by a single person, family, or group of interested persons, from the Latin *imperium*, meaning "supreme or absolute power," especially of an emperor.

SUMERIAN

In 2300 BC Sargon (circa 2334–2279 BC), the ruler of Akkad, conquered the city-states of Sumer (in present-day Iraq) along the banks of the Euphrates River. He created the first empire in history, which lasted until the fall of its last dynasty in 2000 BC.

ROMAN

> *Rome speaks. The case is concluded.*
> —SAINT AUGUSTINE (AD 354–430)

The Roman Empire can trace its origins back to 753 BC after Romulus, who according to legend, killed his twin brother, Remus, to become king. Romulus combined two settlements, one on the Palatine Hill and the other on the Quirinal, to create Rome. The city faced considerable pressure from the Etruscans to the north, who eventually supplanted Romulus with a king of their own.

The city-state of Rome became a republic in 509 BC under the first joint leaders, Lucius Tarquinius Collatinus and Lucius Junius Brutus.

It is generally accepted that the accession to power in 27 BC of Augustus (63 BC–AD 14), the grand-nephew of Julius Caesar (100–44 BC), who had adopted him, marks the de facto beginning of the Roman Empire. Augustus (real name Gaius Octavius Thurinus, also known as Octavian) had himself proclaimed emperor and ruled until his death in AD 14. After years of civil war and instability, Augustus ruled a stable and prosperous empire, with little opposition, even from the urban poor, until his death. The last Roman emperor was Romulus Augustulus who abdicated in AD 476.

PERSIAN

Much of the history of Persia (present-day Iran) is shrouded in the mythology created by successive ancient historians, but it is generally accepted that the Persian Empire began in 559 BC with the rise of Cyrus the Great (580–29 BC), who led an army to topple his grandfather, the tyrannical Hishto-waiga, the king of the Medes. This unified all of the Persian peoples under a single ruler and created the greatest empire the world had known until that time.

Cyrus then led the combined Median and Persian empires to conquer Lydia, which was ruled by King Croesus. The expression "rich as Croesus" originated from the fact that Croesus had coffers overflowing with gold coins. Eventually Cyrus went on to capture Babylon in 539 BC.

The Persian Empire continued to expand, first with the conquest of Egypt and then westward into Europe. Persian expansion was halted by the Greeks at the Battle of Marathon in 490 BC. Herodotus, the great Greek historian, records that 6,400 Persians died and only 192 Greeks.

The last ruler of the remnants of the Persian Empire was Mohammad Reza Pahlavi (1919–80), who was toppled by Ayatollah Khomeini in 1979.

MACEDONIAN

The Macedonian state emerged under its founder Perdiccas I in the seventh or eighth century BC. In the years between 500 and 450 BC Macedonia began to expand into neighboring territories, and with the ascendancy of Philip II (382–36 BC), captured the Greek states of Thracia and Illyria, and the empire began to take shape.

Philip's son, known as Alexander the Great (356–23 BC), captured other Greek states and eventually the Persian Empire, which included part of Egypt. Alexander also added parts of present-day India and Pakistan to his empire.

In 323 BC, at the age of only thirty-three, and with much of the known world at his feet, Alexander fell ill in Babylon and died. The great empire he had so successfully ruled was divided between his generals.

BYZANTINE

The Byzantine Empire was the Greek-speaking, eastern half of the Roman Empire, ruled from Byzantium.

The Roman emperor Diocletian (circa AD 245–312), who ruled from AD 284–305, had split the Roman Empire into eastern and western halves in AD 284 for administrative purposes. Constantine I (known as Constantine the Great) ruled the eastern half until he defeated Licinius, the western ruler, in AD 324, becoming sole emperor of the Romans. He transferred the capital from Rome to the ancient Greek city of Byzantium, situated on the Bosphorus (in present-day Turkey), which he renamed Constantinople (now Istanbul). The Byzantine Empire thus succeeded the Roman Empire. Constantine, the first emperor, ruled from AD 306–37.

Finally conquered by the Ottoman Turks, the Byzantine Empire drew to a close in 1453, almost a full millennium after the fall of Rome in AD 476.

OTTOMAN

The Ottoman Empire began as a small Turkish state around AD 1281 and started to absorb other states from 1451. It was founded by Osman I (hence Ottoman; 1258–1326) and expanded at the expense of the Byzantine Empire, capturing Constantinople in 1453.

It was effectively dissolved in 1923, losing all its territories after the end of the First World War. The Treaty of Lausanne in 1923 established the boundaries of modern Greece and Turkey, which had formed the core of the Ottoman Empire.

HOLY ROMAN

This agglomeration which was called,
and still calls itself the Holy Roman Empire,
was neither holy nor Roman nor an empire.
—VOLTAIRE (1694–1778)

The Holy Roman Empire was the successor to Charlemagne's Christian Empire, which traced its origins from AD 768, the year Charlemagne (AD 747–814) succeeded his father, Pépin III, to become king of the Franks. For the first three years of his reign Charlemagne shared the kingdom of France with his younger brother, Carloman. When Carloman died in 771, Charlemagne became sole ruler and began his Christian imperial expansion.

The period of Charlemagne's time in power is known as the Carolingian Renaissance. According to Carolingian theory, the Roman Empire had only been suspended in 476, and had not ceased to exist. Charlemagne, as the leader of Christianity, was crowned Roman Emperor by Pope Leo III on Christmas Day AD 800.

The first to be crowned as Holy Roman Emperor was Otto I (AD 912–73), who ruled from 962. The empire was partly dissolved in 1648 with the Peace of Westphalia, and the office of Holy Roman Emperor was abolished in 1806.

The Holy Roman Empire was unique in not being based on a nation-state. At its peak it encompassed Belgium, Germany, Austria, Switzerland, the Netherlands, Czechoslovakia, and Slovenia, as well as large parts of France, Italy, and Poland.

MONGOL

The Mongol Empire occupied the largest contiguous geographical area ever to be under the control of a single person. The Mongols never numbered more than 2 million people, but what they lacked in numbers, they made up for in fighting skills, savagery, and determination. They came out of the fringes of the Gobi Desert and struck with ferocious speed to conquer all before them.

The founder of the Mongol Empire, Temuchin, was born around 1167, the son of a local tribal chieftain. Renamed Genghis Khan (Universal Ruler), he united the disparate tribes of Mongol peoples to found the Mongol Empire in 1206.

Genghis Khan used a system of hand signals to control his men, who were organized in decimal units of ten, one hundred, and one thousand, moving them around the battlefield like chess pieces.

By 1214 the Mongol Empire stretched from Poland and Siberia to Vietnam, and from Moscow to the Arabian Peninsula. By 1215 Ghengis Khan controlled most of China. In 1227 he fell from his horse and died.

His grandson Kublai Khan (1219–94) established the Yuan dynasty in China. The dynasty was overthrown in 1368 and the Mongol leaders retreated to modern Mongolia. By 1501 all remnants of the Mongol Empire in China had been destroyed.

SPANISH

In 1469, Isabella of Castile (1451–1504) and Ferdinand of Aragon (1452–1516) married and united their two kingdoms to establish Spain. They began to build an empire, partly to secure their territory against Muslim invaders, but also to protect trade. Their masterstroke was to sponsor the 1492 expedition of the Italian Christopher Columbus (1451–1506) to the Americas. Apart from the earlier acquisition of a few small territories, the Spanish Empire began with the discovery of these overseas lands and all the riches it unearthed.

Under Charles V (1516–22), the Spanish Empire became the most diverse since the Roman, and at its greatest extent, its American territory stretched from Alaska, through western North America and Mexico, to southern Chile and Patagonia. Charles's European lands included Austria, Hungary, and Spain, as well as several Mediterranean islands.

The Spanish Empire came to an end in the War of the Spanish Succession, which lasted from 1701 to 1714. By the end of the hostilities in Europe, Spain had lost Belgium, Luxembourg, Milan, Naples, Sardinia, Minorca, and Gibraltar. The war was fought on several fronts including North America, where it became known as Queen Anne's War. The English effectively captured territory from Florida to Canada, including South Carolina, Massachusetts, Newfoundland, the Hudson Bay region, and St. Kitts.

DUTCH

The Dutch Empire began, as other empires had done, by seeking trading opportunities, which were subsequently defended by force. Its origins are rooted in the success of the Dutch East India Company, which was founded in 1602. The company was granted a twenty-one-year, tax-free monopoly of colonial activities in Asia, and the formation of the company was effectively the beginning of the Dutch Empire. By 1669 the Dutch East India Company had 50,000 employees and an army of 10,000 soldiers. It also possessed 150 merchant ships and 40 warships. It was the biggest private company the world had ever known.

The Dutch Empire continued after the Dutch East India Company ceased to trade in 1798, and by then it encompassed Indonesia, Sri Lanka, Tobago, Surinam, Belgium (1815–30), the Cape Province of South Africa, parts of Malaysia, parts of Brazil, and New York, as well as trading posts over most of the globe. Indonesia, Surinam, and parts of Malaysia remained under the control of Amsterdam until the twentieth century.

In 1624 the Dutch established a colony at Albany in North America with thirty families. Peter Minuit, the governor of the colony, bought an island from the Native Americans and named it New Amsterdam. It is now known as Manhattan.

PORTUGUESE

The Portuguese Empire began in 1415 when Henry the Navigator (1394–1460) captured Ceuta in Morocco from Spain. To finance the expansion of the empire, Portugal focused its energies on trade in gold, spices, and slaves. Portugal eventually acquired vast colonies in Africa, the Far East, China, and India.

The Portuguese explorer Pedro Cabral (1467–1520) discovered Brazil in 1500, and it became a Portuguese colony in 1530 and remained so until its declaration of independence in 1822.

In 1999 Portugal relinquished its final overseas territory when Macao was ceded to China.

FRENCH

In the early sixteenth century the French looked covetously at Portuguese and Spanish empire building.

The earliest French overseas territory, the area around the mouth of the St. Lawrence River in present-day Canada, was discovered by Jacques Cartier (1491–1557) in 1534. He claimed the land for France, named the region New France, and named the high point on an island in the bay Mont Réal (Mount Royal), now Montreal.

The French Empire grew to include the whole of Canada and extended through the Great Lakes, south through Ohio, and down as far as the Mississippi. In 1682, after the French had pushed on and located the mouth of the Mississippi, the whole of Louisiana was claimed for France and named after the French king Louis XIV.

By 1763 all French territory in America except Louisiana had been ceded to the British, and in 1769 the French East India Company, which had been formed in 1664, was dissolved amid massive financial scandals.

A vast area in the center of the continent was sold to the new United States of America in 1803, in what came to be known as the Louisiana

Purchase. The agreed price was $11,250,000, which valued the land at three cents per acre. Louisiana in those days included what would now be one quarter of the continental United States.

There were two French empires. The First Empire, commonly known as the Napoleonic, lasted for a decade from 1804, when the French senate elected Napoléon Bonaparte (1769–1821) as emperor. He abdicated on March 30, 1814, at Fontainebleau, after losing all the territory gained by France since 1792.

The Second Empire covers the period 1852–70. Napoléon III (1808–73) was elected the first emperor, almost unanimously, by plebiscite. During this period, France won territory in Italy, Austria, Syria, Vietnam, and China.

The final remaining French colonies in Africa, among them Algeria, were relinquished amid great bloodshed in the 1950s and 1960s.

GERMAN

The German Empire (also known as the Second Reich) was one of the shortest-lived empires in history. Otto von Bismarck (1815–98) was appointed prime minister of Prussia from 1862–73, and then the first chancellor of Germany from 1871–90. Bismarck allied with Austria to defeat Denmark in 1864, and formed the North German Confederation in 1866.

The German Empire was proclaimed on January 18, 1871, in Versailles after the defeat of the French in the Franco-Prussian War, and King William I of Prussia was proclaimed emperor. As well as its European territories, Germany held Togo, the Cameroons, East Africa, and Southwest Africa (present-day Namibia).

Bismarck resigned in 1890, and Germany lost the whole empire in 1918 after defeat in the Great War (the First World War) of 1914–18.

HAPSBURG

The origin of the Hapsburg Empire (also known as Habsburg) can be traced to the union of the Alpine hereditary lands in Switzerland with the crowns of Hungary, Bohemia, and Croatia in 1526–27. It was dissolved in 1918 at the close of the First World War.

BRITISH

The first reference to a British "empire" was made in 1603 by James I of England (James VI of Scotland) (1566–1625) after he had become monarch of England, Ireland, Scotland, and Wales. The Act of Union of 1707, which united England and Scotland, is generally recognized as the formal beginning of the British Empire.

There have been two distinct periods of British empire, followed by a commonwealth. The first empire was fueled by the commercial aspirations of individuals and organizations, which set out to conquer and exploit overseas lands. The origins of this first empire can be traced to the licensing of chartered companies such as the Muscovy Company, which was formed in 1555. The first empire effectively came to an end with defeat in the American War of Independence, which lasted from 1775 to 1783.

The second empire emerged in the late 1760s and 1770s with Captain Cook's (1728–79) voyages to Australia and New Zealand, and the conquest of most of India under Robert Clive (1725–74) in 1763. At its height, Britain controlled 25 percent of the world's population, and due to its vast geographical extent, it was said that "the sun never sets on the British Empire."

The British Commonwealth of Nations was formed in 1931 to bring together all former dominions and to recognize them as equal and independent nations. This began the demise of the British Empire, which ceded independence to India, Pakistan, and its African territories over the following forty years.

NEW EMPIRES

EUROPEAN UNION

The origins of the European Union (EU) can be traced to a speech given by Winston Churchill on September 19, 1946, advocating a form of

"United States of Europe." He stated, "We need to put the horrors of the past behind us."

Unlike most empires, which formed through conquest, secession, or marriage, the EU, which has been renamed on a number of occasions, has emerged with the full cooperation of the participating countries. This process has been ongoing for more than half a century and continues to the present day.

Main events

The European Recovery Program (also known as the Marshall Plan) was put into effect on June 5, 1947, to stimulate the European economies shattered after the Second World War. Over four years, the United States gave a sum of $13 billion, which would amount to something like £100 billion at the current exchange rate.

The Council of Europe was formed on May 5, 1949. The original members of the council were Belgium, Denmark, France, Ireland, Italy, Luxembourg, the Netherlands, Norway, Sweden, and the United Kingdom.

The European Coal and Steel Community (ECSC) was formed on April 18, 1951. The original members were West Germany, France, Italy, Belgium, the Netherlands, and Luxembourg.

The European Economic Community (EEC) was formed by the same members when the Treaty of Rome was signed on March 25, 1957. The EEC was known in Britain as the Common Market.

Britain was refused membership in the EEC in 1963. The veto was applied with great relish by French president Charles de Gaulle (1890–1970), with his famous and emphatic "Non!" It was always supposed that de Gaulle harbored an old dislike of Britain from the period of the Second World War, when he had been kept out of the planning process for the final assault on mainland Europe, known as D-day.

The European Community (EC) was formed in 1967, when the ECSC, EEC, and EURATOM were merged.

Britain was finally admitted to membership of the EC in 1973 following a national referendum. The EC was renamed the European Union (EU) following the Treaty on European Union or Maastricht Treaty in 1991. The EU now has twenty-five participating countries.

SOVIET UNION

There were two Russian revolutions of 1917, February 24–29 and October 24–25. The Soviet Union was formed after the second revolution.

The first revolution ended with the overthrow of Tsar Nicholas II (1868–1918) and his government, and the installation of a provisional government. The second revolution saw the Bolshevik movement lead the Russian workers and peasants to create the Soviet Union.

Soviet territory was extended to include Armenia, Azerbaijan, Belarus, Estonia, Georgia, Kazakhstan, Kyrgyzstan, Latvia, Lithuania, Moldova, Russia, Tajikistan, Turkmenistan, Ukraine, and Uzbekistan.

In 1956 Soviet tanks rolled into Hungary to impose direct control, and in 1968 Czechoslovakia suffered the same fate. Afghanistan was occupied by the Soviets in 1980.

The disintegration of the Soviet empire began with Mikhail Gorbachev's (b. 1931) reforms in 1985. The empire quickly unraveled with the demand for autonomy by Estonia in 1987, and, by popular demand, it ceased to exist in January 1992.

UNITED STATES

Although there is no U.S. empire in the traditional sense, the United States, with its immense global, political, economic, and cultural influences, has considerable power over the economies of many other countries.

From the thirteen territories that formed the original United States in 1776, the country now comprises fifty states that have been acquired by a variety of means. Some have been purchased from foreign powers (Louisiana from France in 1803 and Alaska from Russia in 1867), some have been ceded to the nation (California from Mexico in 1848), and some were annexed (Texas from Mexico in 1845). Hawaii was annexed voluntarily after a coup financed by American interests.

The United States is the only remaining nation or empire that can successfully intervene as world policeman at its own choice. It is the closest thing to a contemporary empire, and according to the great American writer Gore Vidal (b. 1925), perhaps the last.

CRIME

COVERING PREVENTION AND DETECTION,
ORGANIZED CRIME, PUNISHMENTS,
JUDICIAL SYSTEMS, AND FICTIONAL DETECTIVES

*If this is the way Queen Victoria treats
her criminals, she doesn't deserve any.*
—OSCAR WILDE (1854–1900)

PREVENTION AND DETECTION

U.S. LAW ENFORCEMENT

In the early 1800s the pattern of ethnic diversity changed markedly throughout the original Dutch and English settlements in the United States. A massive influx of German and Irish settlers brought different lifestyles, which were perceived as a threat to the American way of life. Rioting and crime began to flourish. By the middle of the nineteenth century, laws were passed to regulate public behavior and to create penitentiaries and police forces. The model for these newly created police forces was the London Metropolitan Police.

Before America became an independent nation in 1776, and introduced federal laws, the law was enforced on a city-by-city basis, with each city adopting its own legal system. The severity of the punishments relied to a large degree on the prevailing circumstances in the city and the attitude of the city fathers.

The first night watch was established in Boston in 1631, and slave patrols, the forerunners of U.S. police forces, were instituted in the Southern states in 1704.

In the early United States, fines and imprisonment were the main form of punishment for felonies and misdemeanors. Walnut Street Jail in Philadelphia became the first state penitentiary in 1794, with prisoners being segregated for the first time by sex and by the seriousness of their crimes.

U.S. Marshals were first appointed in 1789, under the Federal Judiciary Act, and in 1823 the Texas Rangers were formed to protect settlers from hostile Native American attacks. At the time, Texas was part of Mexico and would remain Mexican until 1836.

The first unified, prevention-oriented police force in the United States was established in New York City in 1845, but uniforms were not issued until 1853.

The Pinkerton Detective Agency was established in 1850 by Allan Pinkerton (1819–84), after success as an amateur in locating the hideout of a counterfeiting gang that had eluded capture by official forces. He was a Scottish barrel maker who had immigrated to America in 1842 and settled close to Chicago.

The forerunner of the FBI (Federal Bureau of Investigation) was a force known as the Special Agents, which was set up in 1908 under the leadership of Attorney General Charles Bonaparte (1851–1921) with just thirty-four agents. At the time, there were only a handful of federal crimes on the statute book, which included banking, antitrust, and land frauds.

The use of 911 as the nationwide emergency telephone number in the United States came about following a request by the National Association of Fire Chiefs in 1957. It was 1968 before a solution was agreed upon and the number 911 was chosen by AT&T. The first 911 call was placed on February 16, 1968, by Rankin Fite, the Alabama Speaker of the House. The call was taken by Rep. Tom Bevill (D), who reportedly answered the call, "Hello." President Clinton signed the bill that designated 911 the nationwide emergency number in 1999, but even today the coverage is not nationwide; some rural areas do not comply with the use of 911.

BRITISH LAW ENFORCEMENT

Scotland Yard (more properly known as New Scotland Yard and sometimes referred to as "The Yard") is the headquarters of the London Metropolitan Police, not the British Police Force. The present building was opened in 1889, and during construction, the dismembered torso of a young woman was discovered. Hence one of the most iconic police headquarters was built on the site of an unsolved murder. The Metropolitan Police was founded in 1829 by Sir Robert Peel (1788–1850).

Fingerprinting to catch criminals was developed in Bengal, India, in 1858 and adopted in England in 1901. Genetic fingerprinting was developed in 1984 by British scientist Alec (later Sir Alec) Jeffreys (b. 1950).

The first use of radiotelephony to apprehend a criminal was in 1910, when Dr. Hawley Harvey Crippen (1862–1910) was arrested as he disembarked from the SS *Montrose* in Quebec after the ship had received a radio signal from England. He was hanged for the murder of his wife, whose dismembered head has never been found.

ORGANIZED CRIME

CHINESE TONGS AND TRIADS

During the 1840s and 1850s the Asian community in the United States, made up principally of Chinese immigrants, experienced racist treatment by whites. In response, the Chinese population in Chicago formed a visible community, and it was in this Chinatown that gang activity first began. Tong gangs were established as merchants' associations, organized to protect members' interests.

Contrary to popular myth, tong gangs are virtually unheard of in China. Triads differ from tongs in having a rigid structure that has existed for hundreds of years in China. Both groups preserve secrecy from the authorities through the concept of "saving face." This code prevents members of the Chinese community from reporting crime, as it would be seen as a failure of the community, which would then suffer loss of face.

MAFIA

The most notorious criminal organization in the world is the Mafia, which originated as a secret organization in Sicily in the late Middle Ages. The Mafia was originally formed to overthrow the rule of foreign conquerors such as the Saracens, Normans, and Spanish.

The Mafia's origins were in small private armies (*mafie*), which began to hire their services to absentee landlords to protect their estates. During the eighteenth and nineteenth centuries these armies became very powerful and began to extort money from their previous employers.

For a time, the Mafia was referred to as the Black Hand Gang, after its signature of a black handprint left at the scene of a murder. This practice fell into disuse after the establishment of fingerprinting.

The Mafia survived successive foreign governments and also the efforts of the Italian dictator Mussolini, who came close to eliminating the organization by the application of repressive methods equally as harsh as those used by the Mafia itself.

After the Second World War the Mafia transferred its powers and interests from rural to industrial areas, and subsequently came to dominate organized crime in the United States.

The first godfather or Capo di Tutti Capi (Boss of Bosses) was "Lucky" Luciano (1896-1962). He created the National Crime Syndicate in America in 1934 through alliances with Meyer Lansky (1902-83) and "Bugsy" Siegel (1906-47), his boyhood friends and fellow gangsters.

Al (Scarface) Capone, perhaps the world's most famous gangster, was never Capo di Tutti Capi of the Mafia in America because he was Neapolitan by extraction, not Sicilian, and therefore not fully trusted.

KRAY AND RICHARDSON GANGS

The Kray and Richardson gangs are the best known of the organized crime gangs that reigned over London during the 1950s and 1960s.

The Kray gang was led by the twin brothers Reggie (1933-2000) and Ronnie (1933-95) with their older brother, Charlie (1927-2000). Their criminal activity began with a protection racket, which they ran from their snooker club in Bethnal Green, London. During the 1950s and 1960s, with a reputation for savage violence, the Krays built a major criminal organization, which vied with the Richardson gang for control of London crime.

The Krays' criminal careers came to an end in 1969 when the twins were convicted of the 1967 murder of Jack "the Hat" McVitie, and were given life sentences. Ronnie died in prison after twenty-eight years, and Reggie was released a few weeks before his death, after serving thirty-two years.

The Richardsons owned scrap yards in south London and operated a lucrative extortion racket at Heathrow car parks, which the Krays tried to take over. The Richardsons' criminal careers came to an end after the "battle of Mr. Smith's Club" in which a cousin of the Krays, Richard Hart, was shot dead.

KU KLUX KLAN

The origin of the name of the Ku Klux Klan (KKK) is uncertain, but one theory maintains that the name comes from the Greek word *kuklos*, meaning "wheel," and *clan*, meaning "family."

The original Ku Klux Klan was a group of six members who met on Christmas Eve 1865 in a law office in Pulaski, Tennessee, supposedly with the intention of forming a social group devoted to wearing weird costumes and playing practical jokes on unsuspecting people. By the end of 1866, in the wake of the American Civil War, its activities had spread beyond Tennessee and progressed to violence. This was aimed at intimidating freedmen, the 4 million former slaves that had been emancipated with the defeat of the Southern states.

In 1867 the Ku Klux Klan appointed its first grand wizard, the former Confederate general Nathan Forrest (1821–77), with the backing of former Confederate leader General Robert E. Lee (1807–70). Some of the other original offices were grand dragon, titan, giant, grand cyclops, and ghoul (the lowest rank).

Up to 1,300 murders were reported during 1868, for which the Ku Klux Klan was blamed. It was disbanded in 1869.

The second Ku Klux Klan was formed in 1915 by William J. Simmons (b. 1880), who set down the prospectus of the organization while he was recovering in a hospital from a road crash injury. Simmons's thinking was strongly influenced by the film *The Birth of a Nation*. The reformation of the Klan was marked by the burning of a cross on Stone Mountain in Atlanta, Georgia, on November 25.

The second version of the KKK is a deeply sinister organization, bent on serious, racially motivated crime. In 2005 it still had approximately 2,500 members.

PUNISHMENTS

Until the late nineteenth century, the usual method of dealing with convicted offenders in Britain was not imprisonment, which is a fairly modern form of punishment, but the imposition of a hefty fine or the meting out of some form of institutionalized brutality. Favorites were public flogging, the ducking stool, mutilation, or branding. Public humiliation was also a favorite, with the use of the stocks or pillory.

The English justice system also imposed transportation, which resulted in extensive numbers of convicted prisoners being taken to America and Australia. Fifty thousand offenders were transported to the American colonies between 1607 and 1776. When these colonies were lost as a result of the War of Independence, Australia became Britain's new penal colony. From 1787, it is estimated that 100,000 men and women were shipped out, before transportation ceased in 1852. During those years, prisoners were held in old ships, the so-called prison hulks. The filthy conditions on board strongly influenced the work of the penal reformer John Howard (b. 1939).

CAPITAL PUNISHMENT

Over the centuries, there have been many methods of capital punishment, such as burial alive, boiling alive, hurling off cliffs, drowning, burning at the stake, crucifixion, decapitation, stoning, hanging, and shooting.

Around 1700 BC, Hammurabi (dates uncertain), the ruler of the Babylonian Empire, codified twenty-five different crimes for which execution was the punishment. Murder was not one of them.

The first recorded death sentence was handed down in Egypt in the sixteenth century BC, when a nobleman was ordered to take his own life, having been found guilty of practicing magic.

Lethal injection

Originally proposed as means of execution by Dr. J. Mount Bleyer of New York in 1888, lethal injection was not used in the United States until Charles Brooks was executed in Texas on December 7, 1982.

Dr. Stanley Deutsch formulated the contents of the lethal injection in 1977.

The Nazis instituted their T4 euthanasia program in 1940 for the mass extermination of hereditarily deformed and mentally disturbed Germans. Phenol (carbolic acid) was injected as the lethal dose.

Electric chair

The original idea for the electric chair came from American dentist Albert Southwick in 1881. Southwick had witnessed the instant death of an elderly drunk who accidentally touched the terminals of an electrical generator.

The first execution by electric chair was of the axe-murderer William Kemmler in 1890 in New York's Auburn State Prison. The alternating current electric chair invented by Dr. J. Mount Bleyer was used, but the whole event was a shambles, with the condemned man suffering a horrible lingering death. The public outcry over Kemmler's death agony played into the hands of Thomas Edison, who had a business manufacturing the far more lethal direct current electric chairs.

Gas chamber

The gas chamber was invented in 1924 by Major Delos A. Turner of the U.S. Army Medical Corps, as a humane way of executing people. It turned out to be the very opposite, since the average dying time from dropping cyanide tablets into acid to produce the poison gas turned out to be more than nine minutes, with several examples of the process lasting twenty minutes.

The first execution by gas chamber took place in Nevada on February 8, 1924. Gee Jon was executed for a murder linked to a tong war. Jon's accomplice, Hughie Sing, was also sentenced to death, but this was commuted to life imprisonment.

Eaton Metal Products of Salt Lake City had almost a complete U.S. monopoly on gas chamber manufacture.

After the prisoner is confirmed dead, by use of a long stethoscope from safely outside the chamber, the warders who enter to remove the dead man or woman are instructed to ruffle the prisoner's hair to release any trapped gas.

Hanging

It was in the fifth century that hanging was first adopted for capital punishment in England. Tyburn, in London, was first used as a place of hanging in 1196 when William Fitz Osbert was made to suffer the noose. The famous Triple Tree of Tyburn was built as a permanent gallows in 1571.

The first execution in America was by hanging. George Kendall, an English colonist, was hung in 1608 for plotting to betray England to the Spanish.

The first woman to be executed in America was Jane Champion; she was hanged at Jamestown, Virginia, in 1632.

Beheading/decapitation

The Bible refers to beheading as a means of capital punishment, although stoning to death was far more common.

The Greeks and Romans used beheading as a less dishonorable method of execution than the alternatives in use at the time. The Romans reserved beheading exclusively for their own citizens, but crucified all others.

William the Conqueror (1028–87) introduced beheading as a means of capital punishment in 1076 but confined it to those of royal birth. The first victim was Waltheof, Earl of Northumberland, who had conspired against the king.

Saudi Arabia beheaded thirty-three men and one woman in 2004.

The Japanese used a particularly brutal form of decapitation until the late 1860s. The victim was buried in the ground and his head was sawn off with a blunt wooden saw.

Guillotine

Widely supposed to have been invented by Joseph-Ignace Guillotin (1738–1814) of France in 1792, the guillotine was intended to provide a more merciful way of beheading people. Because the executioner sometimes failed to behead his victim with the first stroke of the axe, a more reliable method was sought.

In fact Guillotin only went as far as proposing mechanical decapitation. The actual design was by Dr. Antoine Louis (1723–92) of the Paris Academy of Surgery, and the first contraption was built by Tobias Schmidt, a piano maker.

The first execution by guillotine was of the highwayman Nicholas-Jacques Pelletier in April 1792. The crowd, who treated the dance of death in hanging as entertainment, voiced its disappointment and called for the return of the gallows.

Louis XVI, King of France, was guillotined in 1793 on a machine that had been named Louisette.

The first decapitation machine was not the guillotine but the Halifax gibbet, which had been used in Halifax, England, since 1286. John of Dalton was the first to be executed on the gibbet.

The only way to escape the Halifax gibbet was for the condemned person to withdraw his or her head before the blade fell and escape across the parish boundary a mile away. If successful, the felon was then allowed to go free, providing he or she did not return to the town.

A man called Dinnis managed to do precisely that. As he made his escape, he encountered people who were on their way into the town to witness the execution. Asked if Dinnis had been executed yet, he replied, "I trow not," an expression that is still used in the area.

The Halifax gibbet only differed from the guillotine in having a massive horizontal axe blade, whereas the guillotine had a much slimmer blade and a leading edge at forty-five degrees from horizontal. This angle added the "scissoring" effect to the cutting stroke.

In 1564, the 4th Earl of Morton (1525–81) introduced the Halifax gibbet to Scotland, where it became known as the Scottish maiden. On June 2, 1581, the earl himself was executed on the maiden.

The term *gibbet* can mean a gallows on which condemned men and women are hanged, or as in the case of Halifax, a form of beheading machine. In the United States *gibbet* can also mean a device, sometimes body shaped, for displaying the remains of executed criminals to encourage good behavior in others.

Lynching

The extrajudicial execution of lynching refers to the concept of vigilantism in which citizens, in the form of a mob, illegally assume the roles of prosecutor, judge, jury, and executioner.

The term *lynch* is derived from the name of Colonel Charles Lynch (1736–96), a Virginia landowner who began to hold illegal trials in his backyard in 1790. The accused was normally found guilty, then immediately tied to a tree and whipped by Lynch. In some cases he administered the death penalty by hanging.

The first victim in America of this summary justice was John Billington in 1630. He had arrived in America on the *Mayflower* in 1620 as a Pilgrim, and was the prime suspect when his neighbor John Newcomen was shot. He was summarily hanged by a mob of other Pilgrims.

The majority of lynchings in America took place between 1880 and 1930, in most cases the result of racial hatred in the Southern states. Of the 2,800 known victims, 2,500 were black. But according to Ida Wells (1862–1931), an African American journalist born into slavery, writing in the *Chicago Defender*, as many as 10,000 may have been lynched between 1879 and 1898.

The Antilynching Bill was first proposed in 1900 and was passed by the House of Representatives in 1922. Its passage to approval was delayed by a number of filibusters and the Senate finally approved the bill in June 2005 with a formal apology for its failure to approve it earlier.

Prison/jail

In the ancient world, prisons were used for holding the accused pending trial, rather than as a punishment. In ancient Babylonia the state did not act as the prosecuting body, but held accused persons in prison, ready for private prosecutions to be made.

It was not until the late eighteenth century, in England, that imprisonment generally came to be recognized as a punishment at all. Prior to that, prisons had been considered places for holding debtors and prisoners awaiting trial and sentencing.

The first English prison was built in 1557 at St. Bride's Well, London. Subsequently prisons were known as bridewells. There were no individual cells and the whole population of inmates was housed in a single dormitory.

The first prison with cells was built by the Duke of Richmond (1735–1806) in 1775 in Sussex.

The Panopticon prison was designed in 1791 by the English philosopher Jeremy Bentham (1748–1832). It had all the cells on the outside of a circular building, so that all the prisoners could be seen from a central viewing position. The Panopticon was never built in England, but several were built in Europe, Australia, and the United States.

British member of Parliament and social reformer Thomas Fowell Buxton (1786–1845) described a prisoner awaiting trial in 1818:

> The moment he enters prison, irons are hammered on to him; then he is cast into a compound of all that is disgusting and depraved. At night he is locked up in a narrow cell with perhaps half a dozen of the worst thieves in London whose rags are alive and in actual motion with vermin. He may spend his days deprived of free air and wholesome exercise. He may be half starved for want of food, clothing and fuel.

Perhaps the most infamous American prison was Alcatraz in San Francisco Bay in California. Alcatraz Island was originally a military institution, which was established in 1850, and also the site of the first lighthouse on the West Coast. It was converted to a federal penitentiary in August 1934 and held some of the most famous of all prisoners, including Al Capone. Alcatraz closed as a prison in 1963 and was occupied by Native Americans from 1969 to 1971. Today it is one of Golden Gate National Recreation Area's most popular tourist attractions.

The first prison reformer was the Italian Cesare Bonesana, Marchese de Beccaria (1738–94), who campaigned against the use of corporal and capital punishments and torture. Beccaria argued that punishment should have a reforming characteristic and not be used only as vengeance.

Pillory and stocks

The pillory and stocks were common forms of punishment for more than one thousand years before imprisonment was introduced. The victim was displayed in public, either seated in the stocks, secured by the feet, or standing in the pillory, with arms, legs, and head pinioned firmly so that passersby could both view and abuse him or her. The standard procedure was to hurl rotten fruit and eggs at the prisoner, but sometimes harder objects such as stones and rocks were used. Serious injuries and even death resulted from this abuse. The pillory was abolished in England in 1837 but remained in use as whipping posts at three county jails in Delaware in the United States through the 1940s.

A refinement was the revolving pillory, looking rather like a four-fingered road sign, which could hold four prisoners simultaneously, secured only at the head and arms. Each fastening was mounted on a central revolving post, and the criminals were compelled to walk around in a circle, so that they could be viewed from all sides and all could be abused at the same time.

JUDICIAL SYSTEMS

ANCIENT LAW

The oldest laws are held to be those of Ur-Nammu, the founder of a Sumerian dynasty in the city of Ur in Mesopotamia, whose legal code dates from the twenty-first century BC. The code set down laws governing the flight of slaves, punishments for bodily injury and witchcraft, and is written in cuneiform script, the oldest of all writing systems.

JUDGES

In ancient civilizations the earliest legal specialist was the judge. The chief of a society dispensed justice as part of his role of governance, but as his power spread, the chief would delegate the legal function to an official, and often the official sitting as a judge would be a religious man. If not religious himself, the judge would be advised by a religious official.

Judges were first cloaked in robes to follow the fashion of the Roman toga, which was felt to convey authority.

TRIAL BY ORDEAL

The brutal trial by ordeal predates trial by jury. It involved the accused person being required to undertake a painful task, such as carrying red-hot irons. Innocence was established if the accused recovered without injury, on the premise that God would help the innocent.

JURIES

The origins of the jury system are shrouded in mystery, but forms of jury trial are evident in the primitive institutions of most European nations. It is a commonly held, but false, belief that the jury was a creation of Alfred the Great in 870 in England.

The earliest form of trial by jury is the system of "sworn inquest," which was brought to England by the Normans after the conquest in 1066.

The right to trial by jury in the United States was established in 1787 by the first Congress and preserved in the Constitution. The ten amendments, known as the Bill of Rights, were finally ratified by the last state to vote in 1791. The Sixth Amendment states, "An accused person has the right to a speedy trial by an impartial jury."

The first colonial grand jury under the jurisdiction of the British sat in America in 1635.

WORLD'S FIRST DETECTIVE AGENCY

The first known private detective agency was set up in 1833 by Eugène François Vidocq (1775–1857), in Paris. Vidocq called his agency Le Bureau des Renseignments (Office of Intelligence).

After a long career as a criminal, Vidocq managed to persuade the police to employ his services as an informer in exchange for amnesty for his previous crimes. In 1811, after considerable success as a police spy, Vidocq formed the Brigade de Sûreté, the forerunner of the French Sûreté National, and became its first *chef de la police*.

FICTIONAL DETECTIVES

The earliest setting for fictional detectives is the first century BC in ancient Rome. The most prominent example is Gordianus the Finder, created by Steven Saylor (b. 1956). Gordianus is active during the time of well-known figures from history, such as Julius Caesar, Cleopatra, and Mark Antony, and the plots weave real historical characters with fictional ones. Saylor introduced Gordianus in *Roman Blood* (1991), as part of the Roma Sub Rosa series, and set the story in 80 BC. In the story, the young Cicero, a newly qualified advocate, turns to Gordianus for help.

The first modern detective in crime fiction was Dupin. He was the creation of Edgar Allan Poe (1809–49) and was introduced in *The Murders in the Rue Morgue* in 1841, and appeared twice more, in *The Mystery of Marie Roget* in 1842 and *The Purloined Letter* in 1844. Poe created Dupin as the first fictional detective to use superior powers of observation and reasoning to solve crimes and thus paved the way for later characters such as Sherlock Holmes. Poe is regarded as the creator of the modern crime novel.

The most famous fictional detective in the world is probably Sherlock Holmes, introduced in *A Study in Scarlet* in 1887. His creator was Arthur Conan Doyle (1859–1930), who credited his own exceptional powers of deduction to the years he spent studying under Professor John Bell at Edinburgh University.

Conan Doyle was a founder and the first goalkeeper of Portsmouth United Football Club. He also investigated real criminal cases, a direct result of which was the setting up of the Court of Appeal.

Maigret was created by Georges Simenon (1903–89), whose first work featuring Maigret was published in 1935.

Simenon wrote very quickly, expecting to take no more than a month to write a book. Alfred Hitchcock (1899–1980) phoned him one day but was told by his secretary that Simenon couldn't be disturbed because he had just started to write a new novel. Knowing how fast Simenon wrote, Hitchcock said, "Would you mind if I wait?"

Hercule Poirot was first featured by Agatha Christie (1890–1976), the world's most famous crime writer, in *The Mysterious Affair at Styles* in 1920. She wrote this, her first book, while working as a nurse during the First World War.

Miss Jane Marple was another creation of Agatha Christie. *Murder at the Vicarage* was her first appearance in 1930.

Philip Marlowe first appeared in the 1939 novel *The Big Sleep* by Raymond Chandler (1888–1959). In the 1946 film of the book, Humphrey Bogart (1899–1957) provided the definitive portrayal of Marlowe, starring with his future wife Lauren Bacall (b. 1924).

Sam Spade first appeared in 1922 in stories written by Dashiel Hammett in the *Black Mask* magazine. Humphrey Bogart was perfectly cast again in this hard-bitten role in the 1941 film *The Maltese Falcon*.

In the McCarthy "anti-American" witch hunt of the 1950s Hammett refused to give the names of his left-wing friends and was sentenced to six months in prison.

Inspector Morse debuted in Colin Dexter's (b. 1930) novel *Last Bus to Woodstock* in 1975. Morse's forename—Endeavour—was not revealed until 1996, in *Death Is Now My Neighbour*, and he was killed off in the final book, *The Remorseful Day*, in 2000.

DICTIONARIES AND ENCYCLOPEDIAS

COVERING DICTIONARIES, ENCYCLOPEDIAS,
AND *ROGET'S THESAURUS*

DICTIONARIES

The earliest dictionary appeared in the seventh century BC. The Assyrian king Ashurbanipal (circa 669–627 BC) had a dictionary produced on terra-cotta tablets, inscribed with columns of cuneiform writing. The ancient Greeks and Romans both had a form of dictionary in the first century AD. The purpose of these dictionaries was to keep a record of words that had passed out of common parlance, rather than to explain the meaning of words in current usage.

The first English dictionary was *A Table Alphabeticall of Hard Usual Words*, published in 1604 by Robert Cawdrey (b. 1538). The word *dictionary* comes from the Latin word *dictio*, meaning "the art of speaking."

Dr. Samuel Johnson of Lichfield (1709–84) compiled the first comprehensive English dictionary. It was the first dictionary to attempt to standardize the pronunciation and spelling of the English language, and went by the title *A dictionary of the English Language: in which the words are deduced from their originals*.

He began work on his dictionary in 1747, but it was eight more years before he completed it. Despite the dictionary's success, he did not make a significant amount of money from it. Johnson had left Oxford University without a degree but went on to become one of the most quotable men in England.

The *Oxford English Dictionary*, with 15,000 pages, was begun in 1879 and finished in 1928. James A. H. Murray (1837–1915), the largely self-educated first editor, worked on the compilation for thirty-five years but died in 1915 from a heart attack brought on by pleurisy. The dictionary was completed almost exactly seventy years after the Philological Society had first resolved to prepare it.

The first volume of the *OED* appeared in 1888 with the completion of the letter *B.*

Noah Webster (1758–1843) is known as the father of American scholarship and education. Dissatisfied with American primary education, and determined that American schoolchildren should learn from American books, he wrote a three-volume compendium, *A Grammatical Institute of the English Language.* The volumes were published separately: a speller (1783), a grammar (1784), and a reader (1785).

DICTIONARIES OF SLANG

> *Slang is a language that rolls up its sleeves,*
> *spits on its hands and goes to work.*
> —CARL SANDBURG (1878–1967)

The detailed study of slang is recent, with most reference books being produced in the second half of the twentieth century.

The original dictionary of slang, *Vocabulary of the Flash Language*, was written by James Hardy Vaux, a criminal transported from Britain to Australia in 1811. It was published in 1819 and mainly featured the language of the underworld. It was also the first dictionary of any sort to be published in Australia.

An example of the slang is "Frisk the cly and fork the rag, speak to the tattler and hunt the dummy." Translated into normal English, this means: "Pick the pocket and take the money, steal the watch and look hard for the wallet."

ENCYCLOPEDIAS

The first encyclopedia was produced by Plato's nephew, Speusippus, (407–339 BC), who, in about 348 BC, recorded his uncle's ideas on mathematics, natural history, and philosophy. Speusippus also included Aristotle's lecture notes in the encyclopedia.

Vincent of Beauvais (circa 1190–1264), a Dominican friar from Paris, produced *Speculum Maius* (The Great Mirror) in 1244. The book, which ran to more than fifty volumes, claimed to show the whole world not only as it existed, but also as he thought it should be.

The Chinese claim that the *Yongle Canon* or *Yongle Dadian*, compiled between 1403 and 1407 and running to more than 11,000 books, was the world's first encyclopedia. The work was too vast to print and only two manuscripts were made. There are now only four hundred of the original books left, the majority having been destroyed.

The Chinese *Yuhai*, which ran to 240 volumes, was published in 1738.

The prototype of the modern encyclopedia was Ephraim Chambers's *Cyclopaedia*, published in 1728.

The *Encyclopédie* was published in France between 1751 and 1765.

The first English-language encyclopedia was the *Encyclopaedia Britannica*, published in 1768. It became available on the Internet in 1999.

The first multivolume encyclopedia was the *Encyclopedia Americana*. It was published in thirteen volumes between 1829 and 1833 in the United States and by 1919 had expanded to thirty volumes.

Project Gutenberg, which was one of the first free publicly accessible resources on the Internet, went online in 1971 and was the brainchild of Michael Hart (b. 1947) while he was still a student at the University of Illinois. The mission of PG (as it is known) is to digitize, archive, and distribute cultural works, generally in the form of books in the public domain. Hart typed the American Declaration of Independence into the Xerox Sigma Five mainframe computer at the university's Materials Research Laboratory. That same computer became one of the twenty-three that formed the earliest Internet, and the Declaration of Independence, in the words of Hart, "would eventually be an electronic fixture in the computer libraries of 100 million computer users of the future." The earliest attempt to build a free encyclopedia on the Internet can be traced to October 30, 1993, and the original idea of Rick Gates to create Interpedia, one of the predecessors of Wikipedia. (In Hawaiian *wiki wiki* means "superfast.") On March 9, 2000, Nupedia (open source, expert driven) was launched, with Wikipedia following on January 10, 2001.

More than fifty years earlier, Vannevar Bush, Chairman of the National Defense Research Committee, and chief adviser to the president, had proposed the development of hypertext, which he claimed would "lead to new types of encyclopedia."

ROGET'S THESAURUS

This reference book enables writers to find the correct word to express an idea or meaning when that word eludes them. It does not show the meaning of the word or phrase, and it is not arranged alphabetically as are conventional dictionaries and encyclopedias, but according to concepts and themes.

In 1805 at the age of twenty-six, Peter Mark Roget (1778–1869), an English doctor and lecturer of Huguenot descent, began to compile what he called "a classed catalogue of words" to help him express himself better. By 1852, when he was in his seventies, he felt the time had come to publish his great work, which was titled the *Thesaurus of English Words and Phrases.*

Before his death in 1869 there had been twenty-eight further editions of *Roget's Thesaurus*, and right up to the present day it has never been out of print.

ENTERTAINMENT

COVERING ACTING, THEATER, DANCE, HUMOR,
THE CIRCUS, MUSIC HALLS, MUSICALS, FILM,
AND GRAMOPHONE

ACTING

Only since the end of the nineteenth century do we have any photographic, video, or sound recordings of actors' performances. Before then we have to rely on the written recollections of those who witnessed performances, and on paintings that were created to convey how they looked.

Cave paintings in the Trois Frères caves in France dated between 40,000 and 10,000 BC depict men in costume who could conceivably have been acting out a ritual.

However, acting is thought to have begun around 4000 BC, when Egyptian actor-priests worshipped the memory of the dead. It is believed that theater evolved out of primitive rituals, which were created to symbolize natural forces such as fertility or death. By acting, early man may have been reducing these events to a more human scale in an effort to make the unknown more understandable and accessible.

The earliest records of professional nonreligious acting appear in the Sui dynasty of the sixth century BC in China.

THEATER

MIME

Mime is one of the earliest forms of self-expression used by primitive people to convey meaning. Instead of fading from use once spoken language developed, mime changed to become a form of entertainment.

The origins of mime theater are in the fifth and fourth centuries BC at the Greek Theater of Dionysus in Athens, when audiences of ten thousand would watch outdoor festivals in honor of Dionysus, the god of fertility, wine, and theater.

The most elaborate form of mime was hypothesis, which was performed by troupes of actors. The main idea was to develop character rather than plot.

GREEK THEATER

The first time theater is known to have been freed from religious ritual, to become an art form in its own right, was in ancient Greece in the sixth century BC. In Attica, the region surrounding Athens, Greek theater developed from the recital and singing of poetic texts, and from ritual dances to honor Dionysus.

Aeschylus (525–456 BC) introduced the possibility of conflict between characters by introducing a second actor into the dramatic format.

The first known actor was the ancient Greek Thespis, who in the sixth century BC introduced impersonation, or the pretense of being another person. He used masks so that he could play several different characters in the same play.

The first festival of drama was held in Athens in 534 BC, which Thespis is said to have won. Unfortunately no written records of the winners survive. The term *thespian* is used to this day to refer to actors.

ROMAN THEATER

Beginning around 240 BC, Roman theater developed out of the Greek model, but with more emphasis on the voice than was traditional in

Greek plays. The actors, who were mostly slaves, wore masks that may have been designed to amplify their voices to carry to the far reaches of large outdoor theaters. After centuries of competing with the Roman games, in which gladiators fought to the death and Christians were killed by wild animals, Roman theater came to an abrupt end.

PANTOMIME

The ancient Romans, with the approval of Augustus (63 BC–AD 14), developed the *pantomimi*, which was acted out in silence, using only gestures to illustrate the drama and humor. During some early performances of pantomime, Roman audiences thought the actors were Greeks who couldn't speak their language.

PUPPET THEATER

Xenophon (427–355 BC), a Greek soldier and historian, made the first written record of puppet theater in the fifth century BC.

Punch and Judy

The first literary reference to a Punch and Judy show is an entry in the diary of Samuel Pepys for May 9, 1662, in which he writes of having seen what he called an "Italian Puppet Play" in Covent Garden. Pepys recorded that the protagonist was derived from Punchinello, a brutal and vindictive character from the Italian commedia dell'arte, and that Judy was originally called Joan.

MEDIEVAL THEATER

Western medieval drama emerged as an entirely new form, within the church. At Easter and Christmas church services priests would act out small scenes from biblical stories and impersonate biblical figures. These playlets became more elaborate and transferred from inside to the church steps and then to the marketplace. Once the performances were outside church property, artisan guilds took over the responsibility for performances, and nonreligious subject matter was slowly introduced.

Passion plays were first performed in thirteenth-century France and Flanders.

The world-renowned Passion play based on the life of Christ was first performed in 1634. This came about as a result of a vow taken by the residents of Oberammergau (in the Bavarian Alps) to do something for God, in which all the residents, no matter whether rich or poor, would have a part. They hit upon the idea of a play, and they vowed that if only God would spare the town from the spread of bubonic plague, which had already claimed the lives of 15,000 in nearby Munich, they would perform it forever.

The Passion play is performed by the whole community, with 1,700 parts. Performances take place once every ten years and are faithfully continued to this day.

MODERN THEATER

During the early sixteenth century, modern theater originated in Italy. Troupes of actors performed the commedia dell'arte (comedy of art) from town to town. Performance in the commedia dell'arte included improvisation and the invention of new words.

There were ten members in a troupe, and each of them would develop a particular type of character, such as the captain, the valet, the doctor, Harlequin, and so on, with female parts originally being played by men but later by women. Before going on stage they would agree upon a basic plot and a general idea of how it would be performed, and then improvise on stage. The humor was often coarse or bawdy, and rarely subtle.

The Italian troupes traveled all over Europe, strongly influencing theater in Spain, England, Germany, and France.

Method acting

In 1897, Konstantin Stanislavski (1863–1938) and Vladimir Nemirovich-Danchenko (1858–1943) met in a restaurant in Moscow. Eighteen hours and a couple of bottles of vodka later, they had formed the Moscow Art Theater. The mode of acting throughout was to be the Stanislavski system, which became known as method acting, in which the actor's role was to identify himself so thoroughly with the character that he would deliver a naturalistic performance.

Method acting was introduced to America in 1922 when two Russian actors, Maria Ouspenskaya (1876–1949) and Richard Boleslavski

(1889–1937), defected from the visiting Moscow Art Theater. Ouspenskaya founded the School of Dramatic Art in New York and, for the rest of her life, advocated and taught "the method."

DANCE

In its earliest form, dance was not regarded as entertainment (see "Art," p. 18).

The ancient Egyptians in the third millennium BC had formal and ceremonial dances that were part of their religious ceremonies. Eventually the dances became so complex that only specially trained dancers could perform them.

There is evidence that by the fourth century BC dancing had been transformed into something much more personal. There are tales of couples indulging themselves in erotic and lascivious dancing for their own pleasure.

The ancient Greeks were strongly influenced by the earlier Egyptian dances. They held their bull dance on Crete as early as 1400 BC and, using their lively imaginations, developed more than two hundred different dances. Some of these were in preparation for war. They also developed the *kordax*, an uninhibited dance performed in comedies, in which the dancers wore masks.

Plato (428–348 BC) became a dance instructor and wrote texts that divided dance into two types: ugly and beautiful.

The Romans had an early dance of a military nature called the *bellicrepa saltatio*, which is said to have been started by Romulus when he carried off the Sabine women.

Roman dance was courtly and formal, and not universally popular. Men who danced were considered effeminate and all dancing schools were closed in 150 BC.

Romans generally took their public displays seriously and would not allow themselves to look undignified. The Roman statesman and orator Cicero (106–43 BC) said that only madmen danced.

TYPES OF DANCE

The apache is a dramatic dance for a man and woman, with the woman being thrown about the floor by the man. It is said to resemble a pimp dealing with a prostitute and ends with the woman being sent skidding across the floor in an act of violence and submission. There is no connection with the Native American Apache tribe; the dance is named after a Parisian street gang.

The barn dance originated in Scotland around 1860 and was usually held to celebrate the raising of a barn.

The black bottom was introduced in New Orleans in 1919 by songwriter Perry "Mule" Bradford (1893–1970) and the blues singer Alberta Hunter (1895–1984). Bradford claims that he had seen a similar dance performed in 1907 called the Jacksonville rounders dance, which was not a success because *rounder* is another word for a pimp.

The black bottom was a solo "challenge" dance, which featured slapping the backside while simultaneously hopping backward and forward and gyrating the hips. In 1926 the dance became the latest fashion on both sides of the Atlantic, replacing the Charleston.

The cancan emerged in its present form in 1822 in Paris from earlier dances such as the Triori. Risqué rather than lewd, it is a display dance for the stage, which is normally danced to the same tune, the "Galop Infernal" by Jacques Offenbach. It features a troupe of high-kicking female dancers lifting their full skirts to show their underwear and legs, and, to the final notes of the music, doing jump splits.

Until 1866 the cancan was banned from public performance in New York.

The cha-cha, or cha-cha-cha evolved from the mambo. In 1954, after complaints from audiences, Enrique Jorrín (1926–87), a Cuban violinist, slowed down the beat of the mambo to about a third of its original pace to make it easier to perform and thus started the cha-cha.

The dance itself is sensual and sinuous and requires small steps and plenty of hip movement to make it expressive. The name is supposed to have originated from the sound of Cuban ladies' heels tapping the floor in a *cha-cha-cha* sound.

The Charleston was first performed in 1903 in Charleston, South Carolina, and transferred to the stage in Harlem in New York City, in 1913. It was energetic and fun, with plenty of head shaking and ankle lifting.

Harper's Weekly magazine of October 13, 1866, describes a similar dance, which was most likely a version of the twelfth-century French chain dance, the branle.

Clog dancing began in the 1520s in Lancashire, England. It is similar to tap dancing but much slower.

The fandango developed as a seventeenth-century Spanish courtship dance, in which the dancers would dance close together but would not touch. It is said to be the forerunner of all Spanish dances. There is some evidence that a similar dance may have existed in ancient Rome, known even then as the Spanish Dance.

The fox-trot was invented by vaudeville performer Harry Fox (1882– 1959) in 1914. Fox introduced the new dance steps in the Jardin de Danse on the roof of the New York Theater. The fox-trot is unique in being named after the inventor rather than the town or style of dance.

The jive is a vigorous swing dance with plenty of partner twisting and twirling, which developed in the United States in the 1940s.

The jitterbug was created by Cab Calloway (1907–94), the dance band leader, but the word *jitterbug* was actually coined in 1934 by Harry Alexander White. It was a popular exuberant dance in America in the 1930s and 1940s.

During the Second World War, 1,500,000 American servicemen arrived in Britain and brought the jitterbug with them. Horrified dance hall proprietors, not used to seeing such energetic dances and wanting to preserve their dance floors, soon erected notices saying, NO JITTERBUGGING. It had little effect.

Morris dancing is thought to have developed in England in the mid-fifteenth century, although some sources claim it derives from a form of Spanish/English dance current in the 1360s. The term *Morris* is thought to be a corruption of *Moorish* since the Arab or Muslim population of Spain were referred to in English as Moors.

The hokey pokey, or okey cokey originated in Kentucky as a nineteenth-century Shaker song. It is a group dance with silly words; participants stand in a ring during the dance and sing: "You put your right arm in, you put your right arm out . . ." The craze was brought to Britain during the Second World War by American servicemen.

The polka was developed by Joseph Neruba in 1835 in Poland. He had learned the steps from watching a young girl dance and recognized that there was a possibility of making money from the exciting new dance steps.

The tango evolved in the lower-class districts of Buenos Aires in the 1880s. The dance amalgamated elements of the milonga, the mazurka, and the habanera, and by 1900 it had become a worldwide sensation.

Tap dancing developed in the southern United States in the 1830s from dances introduced by African slaves. It also has close connections with clog dancing.

The twist is named for the Hank Ballard (1927–2003) song "Let's Twist Again," which started the craze in 1955, but the dance itself was not popularized until Chubby Checker (b. 1941) released "The Twist" in 1959, opening every performance with a dance lesson. The twist was a worldwide sensation through 1970, and remains a favorite dance of middle-age couples at weddings.

Chubby Checker's real name was Ernest Evans. His stage name was given to him by the wife of Dick Clark as a play on the name of jazz musician Fats Domino.

The waltz originated in Germany in 1520 as a more intimate development of the Westphalia, a thirteenth-century dance. The waltz was one of the first dances in which dancers touched each other, pressing against each other's bodies in public, known as a closed dance. It was at first deemed to be too lewd for women.

HUMOR

Dictionary definition of humor: "Something having the quality of causing amusement or the faculty of expressing or appreciating what is comic or amusing."

—THE PENGUIN ENGLISH DICTIONARY

Comedy is nowadays regarded as a form of drama whose chief objective is to amuse. It was not always so. The Greek philosopher Aristotle (384–322 BC), writing in the fourth century BC, expressed the view that the purpose of comedy is to hold up a mirror to society in the hope that its vices and misbehavior might be corrected. He also states that comedy originated in phallic songs, and that, like tragedy, it began in improvisation.

A century earlier, the Greek comic dramatist, Aristophanes (circa 448–385 BC), wrote around fifty plays, varying in content from savage satire against pretensions to burlesque. His players are the only survivors of "Old Comedy."

Farce is a form of comedy that entertains by means of unlikely and extravagant situations. It is fast-paced, often contains sexual innuendo, and tends to be far-fetched. Farce developed in late nineteenth-century France, with Georges Feydeau (1862–1921) being the preeminent author.

ANCIENT HUMOR

There are twenty-nine "jokes" in the Old Testament, and Jesus (8–4 BC–AD 29–36), like all great public speakers, used a form of humor in his sermons.

Although our modern sense of humor may prevent us from seeing the joke, scholars suggest it is quite likely that Jesus's listeners were doubled up with laughter when they heard Jesus speak about the "speck in your brother's eye" (Matt. 7:3–4), or "pearls before swine" (Matt. 7:6) or the one about "a rich man having as much chance of going to Heaven as a camel passing through the eye of a needle." The "eye of a needle" referred to by Jesus was the common name for a small gate through the wall into Jerusalem. It was passable by pedestrians, but too small for a camel.

It is evident and strangely comforting that Jesus could also take a joke. An early example of irony at his expense came when one of his disciples asked, while looking at him, "How could anything good come out of Nazareth?"

Ancient humor was based on cruelty and harshness until the fifth century BC, when a softer, broader humor began to replace aggression and unpleasantness, and to appear in performance art.

Satire employs humor as criticism to expose something that is foolish or unpleasant. As a form of humor, satire had fallen into decline by the first half of the twentieth century, but was revived on television, almost in its original form, in shows such as *Rowan & Martin's Laugh-In.*

Some Roman writers, such as Quintilian (Marcus Fabius Quintilianus; AD 35–100), claimed that they were the inventors of satire as an anarchic form of drama. But the ancient Greeks, with satirical comedies whose origins can be traced as far back as 500 BC, were writing and performing satires centuries before the Romans.

The oldest surviving satyr play is *Cyclops*, written by Euripides (480–406 BC). The Greek satyr plays with notoriously rude and lacking in subtlety and were copied in Elizabethan times by English writers.

The great Roman philosopher and satirist Horace (Quintus Horatius Flaccus; 65–8 BC) defined the writer of satire as an urbane man who is concerned about the follies he observes, but is moved to laughter rather than rage.

The earliest dramatic comedy emerged from the rowdy choruses and words of the fertility rites of the feasts of the Greek god Dionysus. Old Comedy, as it is known, lasted until around 450 BC and was a collection of loosely connected scenes, which exploited coarse humor in satire, parody, and fantasy, usually at the expense of public figures. Old Comedy was hard-hitting and similar, in a way, to modern satire.

It was superseded by the much milder New Comedy and became less pointedly satirical. The subject matter in New Comedy centered on the exploits of a collection of stock characters such as the thwarted lover, the bragging soldier, and the clever slave. The most famous exponent of New Comedy is Menander (342–391 BC), who was writing in about 320 BC in Greece. One of his few surviving plays is *Dyscolus (The Grouch)* which was rediscovered on a papyrus manuscript in 1957.

EARLY HUMOR

In the Middle Ages, the church tried to keep all the humor and joy out of drama, but it survived in folk plays and festivals.

Court jesters are mainly thought of as existing in the Middle Ages. However, the Roman philosopher and author Pliny the Elder (AD 23–79) mentions court jesters (*planus regium*) in his tale of a visit to King Ptolemy I (367–283 BC).

The earliest entertainers were the *glēomen* (gleemen), who were active in England from the fourth century, entertaining the Angles, a Germanic people who had settled in East Anglia and Northumbria. *Glēomen* would tell amusing stories through song and are described in the Anglo-Saxon epic *Beowulf* as court entertainers. Following the Norman Conquest in 1066 the court and noble house entertainers were known as minstrels, whereas entertainers for the lower classes were known as jongleurs.

The profession of minstrel was at its height between the twelfth and sixteenth centuries, with minstrels being fully employed servants at court as entertainers of any kind, but usually as musicians. A guild of royal minstrels was formed in 1469, but their popularity began to fade during the seventeenth century, and they became obsolete soon after, with the advent of the more sophisticated troubadour.

Troubadours sang of love and chivalry and, as they moved around the kingdom, helped to transmit news from other areas and to spread trade.

MODERN HUMOR
Cartoon characters

Mr. Magoo (first name Quincy) was introduced in 1949 in the short film *Ragtime Bear*. In 1997, Magoo was played by Leslie Nielsen in the film *Mr. Magoo*.

Mickey Mouse, the most famous of all cartoon characters, first appeared in the 1928 film *Steamboat Willie*.

Walt Disney was traveling back by train to his studio in California, having fallen out with his financial backers in New York. They had withdrawn financial support and taken back Oswald the Rabbit, a character that had been copyrighted to them. To fill in the time on the long journey, Disney sketched a new rabbit cartoon character, but changed it to a mouse and called it Mortimer. His wife rechristened the character Mickey Mouse.

Minnie Mouse, the girlfriend of Mickey, also first appeared in *Steamboat Willie*.

Pluto was introduced in 1930 in *The Chain Gang.*

Goofy first appeared in 1932 in *Mickey's Revue.*

Donald Duck put in his first appearance in 1934 in *The Wise Little Hen.*

Bambi, a character from a 1928 book by Felix Salten, scampered onto the screen in 1942.

Li'l Abner was drawn by Al Capp (Alfred Gerald Caplin) (1909–79) for the first time in 1940.

The *Andy Capp* cartoon strip was first published in northern editions of the *Daily Mirror* newspaper in 1957. Capp was created by Reg Smythe (1917–98), who started by submitting cartoons to Cairo magazines while he was stationed in Egypt during the Second World War. *Andy Capp* was syndicated to 1,400 newspapers worldwide and read by 175 million people.

THE CIRCUS

The first building known as a circus was the Circus Maximus in Rome. It is thought to have been founded in the sixth century BC for chariot races. The track was U-shaped with a low wall running down the center and seating on three sides. When it was developed by Julius Caesar (100–44 BC) in the first century BC it could hold 150,000 spectators. It was further extended in the fourth century to seat a quarter of a million people. The Circus Maximus was the largest seated stadium ever built. Nowadays, only the ruins remain.

The modern circus was founded in 1768 by an English trick rider Philip Astley (1742–1814), who found that centrifugal force made it easy to stand on a horse's back while it galloped around in a circle. He set up a ring close to Westminster Bridge, where he operated a riding school in the mornings, and in the afternoons, performed equine acrobatic displays to paying crowds. He took his idea to Paris and formed Le Cirque.

The name *circus* was coined by Charles Hughes (1748–1820), one of Astley's riders, who established his own ring in 1782 and called it the Royal Circus. Hughes took his ideas, which by then incorporated jugglers, trapeze artists, clowns, and animals, to Russia, and the Moscow State Circus was born.

The first circus tent was introduced by the American J. Purdy Brown in 1825. Before the advent of the "big top," circuses had been held either in permanent buildings or outdoors.

MUSIC HALLS

Music-hall songs provide the dull with wit,
just as proverbs provide them with wisdom.
—W. SOMERSET MAUGHAM (1874–1965)

This form of public entertainment developed in the eighteenth century from "taproom concerts," which were popular in English city taverns. A number of acts of Parliament were passed to restrict music and entertainment in taverns. These had precisely the opposite effect and resulted in the rapid development of specially constructed halls, where smoking and drinking were permitted, to accommodate the demand.

The first dedicated music hall was built in 1852 by Charles Morton (1819–1904). He called it Morton's Canterbury Hall, and featured acts were Harry Lauder (1870–1950), Dan Leno (1860–1904), and Vesta Tilley (1864–1952), the biggest names of the time. As a form of popular entertainment, the music hall went into decline in the 1920s with the advent of "talking pictures." Several halls were converted into cinemas.

MUSICALS

COMIC OPERA

Modern musicals are descended from comic operas such as *The Beggar's Opera* by John Gay (1685–1732), which debuted in 1728. Composers of comic operas borrowed popular songs of the day and rewrote the lyrics to suit the plot. Tunes by Handel and Purcell were commonly used. Robert Walpole (1676–1745), who was the prime minister of the day, and the butt of Gay's satire in *The Beggar's Opera*, arranged for Gay's next opera, *Polly*, to be banned.

OPERETTA

The first operetta, which bridged the gap between opera (see "Art," p. 11) and the musical, was *Orpheus in the Underworld*, composed by Jacques Offenbach (1819–80) and first performed in 1858. As a German composer living in Paris, Offenbach experienced huge difficulties because opera was a government monopoly in France, and he was seen as the main competition.

MINSTREL SHOWS

The pioneers of minstrel shows as we think of them today were Dan Emmett's Virginia Minstrels. In 1843, Emmett (1815–1904) staged an African American spoof version of the Tyrolese Minstrel Family, who toured the United States from 1840, with a series of musical comedy sketches that featured fake European music. He offered full shows of white performers with "blacked up" faces, performing song and dance routines.

African American impersonation shows declined in popularity toward the end of the nineteenth century but enjoyed a brief revival on British television in the 1950s and 1960s. However, this form of entertainment died out as society grew to regard "blacking up" as offensive.

MUSICALS AND MUSICAL COMEDY

The first modern musical is believed to be *The Black Crook*, which opened at Niblo's Gardens in New York City on September 12, 1866. The production was based on a book by Charles Barras.

The first musical comedy was *The Mulligan Guard* in 1878. A uniquely American creation, it packed the Theatre Comique on Broadway for the whole month of its run. It was produced by and starred Ned Harrigan and Tony Hart.

The first British musical was *Dorothy*. It was staged by George Edwardes, the theater impresario, with music by Alfred Cellier at the Gaiety Theatre, London, in 1865. *Dorothy* was a three-act musical, which broke box office records by playing for 931 performances, after transferring to the Prince of Wales Theater later that year.

The first Hollywood musical was *The Jazz Singer*, in 1927, starring Al Jolson (1886–1950). *The Jazz Singer* was also the first feature film to use sound successfully. Jolson's second Hollywood musical, *The Singing Fool* of 1928, set box office records that lasted until *Gone With the Wind* in 1939.

FILM

The appearance of motion when a series of still photographs are projected rapidly onto a screen, as experienced when watching a film, is made possible by an optical phenomenon known as persistence of vision. The main technical innovations involved in producing moving pictures were achieved in France in the 1880s and 1890s by brothers Auguste (1862–1954) and Louis (1864–1948) Lumière, and in the United States by Thomas Alva Edison (1847–1931).

However, a groundbreaking contribution was made in 1889 by William Friese-Green (1855–1921) of England, who invented the cine camera. He recorded a film of the Esplanade in Brighton using paper film negative. Later the same year he replaced the paper with celluloid, which remains the standard material to this day.

The first public screening of celluloid film was hosted by the Lumière brothers on March 22, 1895.

The first international film empire was created by Charles Pathé (1863–1957) of France. The Pathé name is still familiar to modern filmgoers.

The first Hollywood film studio was opened by the Nestor Company in 1911 in the converted Blondeau Tavern. Cecil B. DeMille (1881–1959) and Sam Goldwyn (1879–1974) soon followed, and fifteen other film companies were established within a year. The Hollywood film industry grew rapidly to dominate the world of motion pictures.

In 1886 H. H. Wilcox bought an area of land on Rancho La Brea to develop as a residential community for the wealthy. His wife, Daeida, rechristened the area Hollywood after hearing the name on a train journey. Wilcox paved a road, which he called Prospect Avenue. It became one of the world's most famous streets, known now as Hollywood Boulevard.

The first full-length feature film was *The Squaw Man*, produced in 1914 by DeMille and Goldwyn, who went on, in 1917, to establish a hallowed Hollywood tradition with the first sequel, *The Squaw Man's Son.*

The first Hollywood superstar was Mary Pickford (1892–1979) who, in 1927, cofounded the Academy of Motion Pictures, which awards the Oscars annually.

Most of the great film studios originated in this era:

- Paramount Pictures in 1912
- Columbia Pictures in 1920
- Warner Brothers in 1923
- MGM in 1924

United Artists was formed as a partnership in 1919 between Charlie Chaplin (1889–1977), Douglas Fairbanks (1883–1939), and Mary Pickford (1892–1979), together with director D. W. Griffith (1875–1948). The other studios could not afford their salaries.

COLOR FILM

The first color film process was Charles Urban's Kinemacolor, which was produced in 1906. Its huge cost prevented it being a commercial success.

Technicolor was developed by German engineer Helmut Kalmus in 1922, and became the industry standard.

SOUND FILM

The first system to attempt to put sound on film was the Gaumont Chronophone system, which was launched in 1900, but it failed to synchronize sound and vision well enough.

The first usable sound system was the Phono-Bio-Tableaux of 1905, which was developed by Walter Gibbons. The first performer to be heard on the new system was Vesta Tilley (1864–1952) in the same year, but it would be 1927 before the first "talkie," *The Jazz Singer* with Al Jolson (1886–1950), would fully incorporate sound.

CINEMASCOPE

A physicist, Henri Chrétien (1879–1956), invented CinemaScope in 1928. It is a wide-screen film format and was largely ignored until the 1950s, when competition from television as the most popular form of entertainment forced film companies to offer a better film experience to their audiences.

The first CinemaScope film was *The Robe*, released in 1953.

FILM RATINGS

The Motion Pictures Producers and Distributors Association (MPPDA) was formed by American studio bosses in 1922 in response to a number of scandals, mainly of a sexual nature, which had rocked the industry. The association's objective was to introduce a system of self-regulation into the production of movies, and they hired Will H. Hays (1879–1954), a former U.S. postmaster general, as president. Hays had masterminded Warren Harding's successful campaign for the U.S. presidency and was known as a man of unimpeachable, if rather stolid, character.

By 1930 the so-called Hays Code (also known as the Production Code) was introduced, but there was no effective means of enforcement. After July 1, 1934, all films required a certificate of approval before release, which meant the Hays Code could now be properly applied. In 1968, the Motion Picture Association of America (MPAA) succeeded the MPPDA to fulfill the movie industry's self-prescribed obligation to the "parents of America."

On November 1, 1968, the new code introduced ratings under the following labels: **G** (suitable for general audiences of all ages), **M** (suitable for mature audiences; parental guidance suggested, but all ages admitted), **R** (restricted: no one under sixteen admitted unless accompanied by an adult), **X** (no one under seventeen admitted).

FILM FESTIVALS

Usually, film festivals are held annually, for the purpose of evaluating recently released, or shortly to be released, motion pictures. Critics, distributors, filmmakers, and others with an interest in the promotion of films normally attend. Side attractions include the opportunity for the general public to see film stars close at hand and to sample the glamour of such occasions. It is normal practice for the paparazzi to cluster in any likely spot, such as a hotel pool, where a starlet may be photographed wearing as little as possible.

The first film festival was held in 1932 in Venice, and has been an annual event ever since. Major festivals are now held in Berlin, Cannes, Moscow, London, New York, and San Francisco.

THE OSCARS

More properly known as the Academy Awards, the Oscars were first awarded in 1929. The prizes were small golden statues and were first referred to as Oscars in 1931 by Margaret Herrick, a director of the Academy, who thought the statuette resembled her uncle Oscar. The name stuck.

The winners of the first Oscars were:

Best actor (lead) Emil Jannings (1884–1950) for *The Last Command* and *The Way of All Flesh*. Unlike today, when the suspense is maintained right to the last moment, in 1929 the winner was known several days before the event. Jannings, who was German, had booked to return to Europe and asked to be presented with the award ahead of the actual ceremony. His wish was granted, and he became the first ever recipient of an Academy Award.

Best actress (lead) Janet Gaynor (1906–84) for *Seventh Heaven*.

Best director Frank Borzage (1893–1962) for *Seventh Heaven*.

Best picture *Wings*.

The first family to produce three Oscar winners was the Hustons: Walter (1884–1950), John (1906–87), and Anjelica (b. 1951).

The first British winner was George Arliss (1868–1946), who played British prime minister Benjamin Disraeli in the 1929 film *Disraeli*. Arliss had specialized in portraying Disraeli in his one-man stage show and was regarded as a "natural" for the part.

The first use of the sealed envelope for announcing winners was in 1941.

The first African American winner of the best leading actor award was Sidney Poitier (b. 1927), in 1963 for *Lilies of the Field.*

The first African American actress to win the best leading actress award was Halle Berry (b. 1966), in 2002 for *Monster's Ball.*

The first African American to be awarded an Oscar in any discipline was Hattie McDaniel (1895–1952), who won the best supporting actress award in 1939 for her portrayal of Mammy in *Gone With the Wind.*

The first radio broadcast of the Oscars was in 1944.

The first televised broadcast of the Oscar ceremony was in 1953, and 1966 was the first year the ceremony was broadcast in color.

GRAMOPHONE

Thomas Alva Edison (1847–1931), the great American inventor, produced the forerunner of the gramophone, the phonograph, in 1877.

Emile Berliner (1851–1929) coined the word *gramophone* in 1894 as a trademark for his new and improved design of phonograph. Berliner's machine used flat discs in place of the cylinders that were used in phonographs.

JUKEBOXES

The name *jukebox* may have derived from the Elizabethan word *jouk*, meaning "to move quickly or dodge," or the African American slang term *jook* meaning "to dance."

The first coin-operated machine to play recorded music was installed on November 23, 1889, by Louis Glass (1864–1936) in the Palais Royal saloon in San Francisco. He linked an electrically operated Edison phonograph to four listening tubes, each of which was separately coin-operated. The phrase "nickel in the slot machine" was coined to describe it.

The first jukebox as we would recognize it today was invented in

1905 by John C. Dunton of Grand Rapids, Michigan. The customer had a choice of twenty-four preselected recordings, each of which was on a cylinder.

Jukeboxes went into decline during the 1930s and 1940s until demand was stimulated by better sound quality and the advent of rock-and-roll music in the 1950s.

HIT PARADE

The first sheet music sales chart was topped by the Ink Spots in January 1950 with "You're Breaking My Heart."

The original number one on the U.S. *Billboard* **"Hot 100"** chart was "Wheel of Fortune" sung by Kay Starr (b. 1922) in March 1952.

In 1949 Todd Stortz bought the ailing radio station KOWH in Omaha, Nebraska, and revitalized its fortunes by pioneering the radio format of playing the Top 40 hits in a countdown to number one.

Radio Luxembourg adopted the same format and broadcast the Top 20 throughout the 1950s and 1960s.

The original British number one hit in the record sales chart was "Here in My Heart" by Al Martino (b. 1927), in October 1952. As of 2006 Al is still singing for his living.

COMPACT DISC (CD)

The CD, a molded plastic disc for the reproduction of sound, was developed by Philips Industries (the Netherlands) and demonstrated in 1980 at the Salzburg Festival by Herbert von Karajan (1908–89), the lead conductor of the Berlin Philharmonic Orchestra.

At the launch event, von Karajan said, "All else is just gaslight," a reference to the fact that CDs were the recorded music format of the future. The prototype CDs were designed for sixty minutes of music. It is rumored that von Karajan used his influence with Philips to promote the longer seventy-four-minute CD, in order to accommodate Beethoven's Ninth Symphony, which was his favorite piece of music.

Philips and Sony (Japan) worked together on more than one hundred classical and jazz recordings, and launched CDs into the recorded music market in 1982.

DIGITAL VERSATILE DISC (DVD)

In 1995 Philips/Sony demonstrated its new format MMCD (multimedia CD). In the same year Toshiba/Warner demonstrated its new format SD (super disc). The two consortia agreed to combine the best of both formats to produce the single standard-format DVD.

DVD (ROM) and DVD (Video) were launched in Tokyo in November 1996, and in the United States in August 1997. The first DVD was sold in Europe in 1998.

ELECTRIC GUITAR

The electric guitar was crucial to the development of rock and pop music. Lloyd Loar of the United States developed the original prototype in 1924, and his first commercial model, the Vivi-Tone, was launched in 1933. It was a flop.

MOOG SYNTHESIZER

Robert Moog (1934–2005), an American engineer, invented the synthesizer in 1964. He developed his ideas from building theremin kits at home. (The theremin was named after Léon Theremin (born Lev Sergeivitch Termen; 1896–1993), a Russian cellist, who, in 1917, developed an electronic machine that could make music by combining two different high-frequency sound waves. Soviet leader Vladimir Ilyich Lenin owned a theremin and had lessons on how to play it from Theremin himself.) Moog sold around one thousand of these before starting to make his own instruments, collaborating with two composers, Herbert Deutsch and Walter (later Wendy) Carlos (b. 1939).

After the success of Carlos's album *Switched on Bach*, which was the first record produced using only electronic instruments, the Beatles and Rolling Stones bought synthesizers and the name Moog passed into pop legend. After his synthesizer company went bankrupt, Robert Moog returned to producing theremins.

Moog is one of the most mispronounced words in music. It should be pronounced to rhyme with *vogue*.

POP MUSIC

The "big bang" year in pop/rock music was 1954. Elvis Presley released "That's Alright Mama" with Sun Studios, and Bill Haley released "Shake, Rattle and Roll."

Pop stars, their first number one hits

Bill Haley and the Comets (previously the Saddlemen), "Shake Rattle and Roll," 1954 (USA).

Little Richard, "Tutti Frutti," 1955 (USA).

Elvis Presley, "Heartbreak Hotel," 1956 (USA).

Buddy Holly (with the Crickets), "That'll Be the Day," 1957 (USA).

Cliff Richard, "Living Doll," 1959 (UK).

The Beatles, "Please Please Me," 1962 (UK), or depending on which UK chart compiler is believed, "From Me to You," 1963 (UK).

The first recording made by John Lennon, George Harrison, and Paul McCartney was "If You'll Be True to Me."

The Beach Boys, "I Get Around," 1964 (USA).

Michael Jackson, "Ben," 1972 (USA).

Jackson was a member, along with four of his brothers, of the Jackson Five, which had a number one hit with "I Want You Back" in 1970.

The Rolling Stones, "It's All Over Now," 1964 (UK).

Note: there is still some debate over the details of some of these entries.

The first album to reach one million in sales

The soundtrack to *The Sound of Music*, in 1963.

Groups—their original names

Famous Name	Original Name
Blondie	Angel and the Snake
Creedence Clearwater Revival	The Blue Velvets
Bill Haley and the Comets	Bill Haley and His Saddlemen
The Beach Boys	Carl and the Passions
The Supremes	The Primettes
Simon and Garfunkel	Tom and Jerry
Sonny and Cher	Caesar and Cleo
Black Sabbath	Polka Tulk

Mamas and Papas	The New Journeymen
The Temptations	The Elgins
The Beatles	Johnny and the Moondogs
	(also the Quarrymen)
Buddy Holly	Charles Hardin Holley
Cliff (later Sir Cliff) Richard	Harry Webb

FAMOUS PEOPLE

COVERING U.S. PRESIDENTS, OTHER WORLD LEADERS,
THE ANCIENTS, ACTORS AND OTHER
SHOW BUSINESS PEOPLE, WRITERS, SINGERS,
GREAT ARTISTS, AND OTHERS

The famous exist in the public gaze, and the man in the street wonders: How did they get there? Where did they spring from? What did he do before he became the pope or a film star?

What did Hitler do to make ends meet before he ran Germany? Was there something odd about Julius Caesar? Did Sean Connery have a "proper job" before he started acting?

Public images hide as much as they reveal, and what lies hidden is the most intriguing of all. What you see is not always what you get. Some famous people change their birth names for show biz or political reasons. Alternatively, a new name might be adopted just because the real name is so awful. Imagine, for example, what the parents of the Hollywood film star John Wayne were thinking by naming him Marion. Why would the great Chinese leader Chiang Kai-shek need four names at different times of his life? You may also be surprised to learn that George Washington was once an officer in the British Army, and one of the best-known Hollywood "tough guy" stars began his career as a female impersonator. Perhaps not so difficult to imagine is that a recent British prime minister failed to qualify as a bus conductor, went on to work in a bank, and then achieved the highest office in the land. Before they became famous, one star worked as a zoo attendant, one a concert pianist, and three worked for the U.S. Postal Service.

This chapter reveals the surprising or little known origins, or backgrounds, of famous people. It could be their original names or the careers they pursued before the one that made them famous. It could be a snippet of information that just completes that ever-elusive picture.

U.S. PRESIDENTS

George Washington (1732–99)
Tobacco farmer
As a lieutenant colonel in the British Army he surrendered to the French at Fort Duquesne.

Thomas Jefferson (1743–1826)
Lawyer
Planter

Abraham Lincoln (1809–65)
Mill manager
Postmaster
Lawyer

Harry Truman (1884–1972)
Farmer
Lead mine owner (failed)
Oil prospector (failed)
Haberdashery shopkeeper (failed)
Partner in a bank (failed)
County court judge
Truman entered the U.S. Senate in 1935. He became president when Franklin Roosevelt died toward the end of the Second World War. Truman had met Roosevelt only twice and took office with almost no knowledge of the current Second World War plans.
"I never give the general public hell.
I just tell them the truth and they think it's hell."

John F. Kennedy (1917–63)
Inherited colossal wealth from his father's liquor
smuggling during Prohibition
Served in the U.S. Navy in the Second World War
                "We will go to the moon."

Lyndon Johnson (1908–73)
Schoolteacher
National youth administrator

Richard Nixon (1913–94)
Performed in amateur theater
Liutentant commander in U.S. Navy
Applied and failed to become an FBI agent
Lawyer
The only man to have been elected to two terms as vice
president and two terms as president.
        "People have got to know whether their president
                is a crook. Well I'm not a crook."

Gerald Ford (b. 1913)
Football player
Lieutenant commander in U.S. Navy

Jimmy Carter (b. 1924)
Peanut farmer
U.S. Navy officer

Ronald Reagan (1911–2004)
Film actor
Union boss
            "You can tell a lot about a fellow's character
                by his way of eating jelly beans."

George H. W. Bush (b. 1924)
Fighter pilot in the Korean War (bailed out after aircraft
was hit)

Bill Clinton (b. 1946)
University teacher

"I'm not going to say this again. I did not have
sexual relations with that woman,
Miss Lewinski. These allegations are false."

George W. Bush (b. 1946)
Oilman
F-102 pilot in the Texas Air National Guard
Part owner Texas Rangers baseball team

OTHER WORLD LEADERS

Winston Churchill (1874-1965)
Descendant of Duke of Marlborough
Half American on mother's side
War reporter in the Boer War
Took part in the last British Army cavalry charge, under
Lord Kitchener

Harold Wilson (1916–95)
Contracted typhoid in 1931
University lecturer
Member of Liberal Party before joining Labour
President of the Board of Trade at age thirty
"In politics, a week is a very long time."

James Callaghan (1912–2005)
Tax inspector
Lieutenant in the Royal Navy
Callaghan is the only man to have held all three great
offices of state before becoming British prime minister.

Margaret Thatcher (b. 1925)
Research chemist at British Xylonite
Food chemist at J. Lyons, developed soft ice cream
Lawyer

John Major (b. 1943)
Bus conductor (failed)
Garden gnome manufacturer
Bank clerk

Tony Blair (b. 1953)
Father became a Communist, later a Tory
Rock band guitarist
Lawyer
Pat Phoenix, a leading actress in Coronation Street, campaigned for Blair in 1983.

Yasser Arafat (1929–2004)
Studied civil engineering at the University of Texas
Qualified as a civil engineer and became a businessman

Golda Meir (1898–1978)
First female prime minister of Israel
Described by David Ben-Gurion, the founder of modern Israel, and its first prime minister, as "the only man in the cabinet." Meir went to school in Milwaukee from 1906 to 1914 and re-immigrated to Israel in 1921.

Sirimavo Bandaranaike (1916–2000)
World's first female prime minister
Prime minister of Sri Lanka 1960–65, 1970–77, and 1994–2000
Bandaranaike's husband, her predecessor as prime minister, was assassinated by a Buddhist monk in 1959.

Edith Cresson (b. 1934)
First female prime minister of France (1991–92)
Considered borderline racist, accusing the Japanese of being "ants trying to take over the world," and homophobic, referring to homosexuality as an "Anglo-Saxon problem."

Pope Benedict XVI (b. 1927)
Real name: Joseph Alois Ratzinger
Member of the Hitler Youth

German army deserter
Allied prisoner of war
Professor at several German universities

Julius Caesar (100–44 BC)
Full name: Gaius Julius Caesar
Suffered from epilepsy
Studied oratory in Greece
Slept with Pompey's wife

Mahatma Gandhi (1869–1948)
Real name: Mohandas Karamchand Gandhi
Married at thirteen
Barrister in England
Political activist in South Africa
Survived a lynching attempt

Chiang Kai-shek (1887–1975)
Given name at birth: Chung-Cheng (which means
Balanced Justice)
Milk name: Jui-Yuan (which means Auspicious Beginning)
Pinyin: Jiang Jie-shī (which means Between the Rocks)

Lenin (1870–1924)
Real name: Vladimir Ilyich Ulyanov
Born into a middle-class family
   "Liberty is precious. So precious it must be rationed."

Pope John Paul II (1920–2005)
Real name: Karol Józef Wojtyla (pronounced
Voyteelah)

Joseph Stalin (1878–1953)
Real name: Iosif Vissarionovich Dzhugashvili
Studied for the priesthood
Bank robber

Leon Trotsky (1879–1940)
Real name: Lev Davidovich Bronstein
Murdered in Mexico by Stalin's agent Ramon Mercader

THE ANCIENTS

Pythagoras (582–07 BC)
Famous for his theorem "the square of the hypotenuse is
equal to the sum of the squares of the opposite two sides."
He also invented the eight-note music scale, having heard
the different "notes" emanate from unequally sized anvils
being struck by blacksmiths.

Aristotle (384–22 BC)
Private tutor to Alexander the Great

Plato (circa 427–347 BC)
Real name: Aristocles
Amateur wrestler
Nicknamed Plato because his wide shoulders gave him a
"platelike" appearance.

ACTORS AND OTHER
SHOW BUSINESS PEOPLE

AMERICAN

Woody Allen (b. 1935)
Newspaper joke writer

Pamela Anderson (b. 1967)
Nude calendar model for *Playboy*

Louis Armstrong (1901–71)
Nickname: Satchmo (short for "satchel mouth")

Fred Astaire (1899–1987)
Real name: Frederick Austerlitz

Lauren Bacall (b. 1924)
Real name: Betty Joan Perske
Related to Shimon Peres, former Israeli prime minister

Humphrey Bogart (1899–1957)
Real name: Humphrey DeForest Bogart
Son of a famous New York surgeon and a margarine
demonstrator
The distinctive feature of his set upper lip was the result of
a shell wound received in the Second World War.

James Cagney (1899–1986)
Half Norwegian on mother's side
Began stage career as a female impersonator

Bing Crosby (1903–77)
Real name: Harry Lillis Crosby
Adopted Bing from his favorite comic strip, *The Bingville
Bugle*
Postal clerk

James Dean (1931–55)
First appearance on film was in a Pepsi-Cola
advertisement
Last appearance on film was in a road safety
advertisement (he died in a motor accident days later)

Cecil B. DeMille (1881–1959)
Owned a girls' school

Walt Disney (1901–66)
Assistant post office letter carrier

Robert Downey Jr. (b. 1965)
Shoe salesman

Richard Dreyfuss (b. 1947)
Conscientious objector during the Vietnam War

Robert Duvall (b. 1931)
Son of a U.S. Navy admiral
Descended from Civil War general Robert E. Lee

Clint Eastwood (b. 1930)
Firefighter
Lumberjack
Steel-mill furnace stoker
Swimming instructor and lifeguard

Peter Falk (b. 1927)
Lost right eye at the age of three from a malignant tumor
Worked as an efficiency expert for the Connecticut State
Budget Department

Henry Fonda (1905–82)
Journalist

Harrison Ford (b. 1942)
Carpenter

Jodie Foster (b. 1962)
Real name: Alicia Christian Foster

Judy Garland (1922–69)
Real name: Frances Gumm

Cary Grant (1904–86)
Real name: Archibald Leach
Born in Bristol, England
Acrobat
Song and dance man
Stilt walker
Juggler
Lifeguard

Dustin Hoffman (b. 1937)
Janitor
Attendant in mental hospital

Bob Hope (1903–2003)
Real name: Leslie Townes Hope
Born in Eltham, England

Rock Hudson (1925–85)
Real name: Roy Scherer
Post office letter carrier
In 1985 Hudson became the first famous person to die of
AIDS.

Stan Laurel (of Laurel and Hardy) (1890–1965)
Real name: Arthur Stanley Jefferson
Music hall and American vaudeville

Jack Lemmon (1925–2001)
Pianist in a beer hall

Jayne Mansfield (1933–67)
Real name: Vera Palmer
Concert pianist and violinist

Lee Marvin (1924–87)
Plumber
U.S. Marine
Marvin became the world's first palimony case after
splitting with his live-in girlfriend of six years.

Robert Mitchum (1917–97)
Heavyweight boxer
Aircraft fitter
Stagehand

Marilyn Monroe (1926–62)
Real name: Norma Jean Baker (or Norma Jean
Mortenson)
Nude calendar model on the cover of the first issue of
Playboy

Paul Newman (b. 1925)
Second World War naval radioman

Jack Nicholson (b. 1937)
Office boy in MGM cartoon department

Al Pacino (b. 1940)
Theater usher
Porter
Superintendent of an office building

Gregory Peck (1916–2003)
Medical student

Sidney Poitier (b. 1927)
Physiotherapist

Anthony Quinn (1915–2001)
Irish-Mexican son-in-law of Cecil B. DeMille

Robert Redford (b. 1936)
Pavement artist in Paris at age nineteen

Martin Sheen (b. 1940)
Real name: Ramon Estevez
Janitor
Soda jerk
Half Irish, half Spanish
Mother fled Ireland during Irish War of Independence
due to IRA connections

Kevin Spacey (b. 1959)
Real name: Kevin Fowler
Stand-up comedian in comedy clubs

Sylvester Stallone (b. 1946)
Expelled from fourteen schools in eleven years
Zoo attendant
Porn film actor (*The Party at Kitty and Stud's* and *Italian Stallion*, 1970)
Pizza demonstrator
Theater usher

Barbra Streisand (b. 1942)
Switchboard operator
Theater usher

John Wayne (1907–79)
Real name: Marion Morrison

Mae West (1893–1980)
During the Second World War an inflatable yellow life
jacket was named after her.

Louise Brooks (1906–85)
Victim of serious sexual abuse as a child
Popularized the "bob" haircut
Dancer with the Ziegfeld Follies
Was the inspiration for *Dixie Duggan*, a newspaper strip
that ran from the 1920s until 1962

Halle Berry (b. 1966)
Named after Halle's Department Store in Cleveland, Ohio
Half British (mother is from Liverpool)
Placed sixth in Miss World 1986

Elizabeth Taylor (b. 1932)
Born in Hampstead, London
Appeared in her first film at age nine
On her father's side, Taylor is descended from King
Malcolm II of Scotland, who reigned from 1005 to 1034.

Bess Flowers (1898–1924)
Appeared in more films than any other actor, male or
female (career span 1923–64)
Known as the Queen of Hollywood Extras, she is the only
extra to receive a credit in the *Film Encyclopedia*.

Natalie Wood (1938–81)
Real name: Natalia Zakharenko, became Natasha Gurdin
White Russian parents
Had a deeply held premonition of her own drowning in
deep water, which had been forecast by her mother

Julia Roberts (b. 1967)
While her mother was pregnant with Julia, her parents ran an acting school that was attended by Martin Luther King's children.

Kim Basinger (b. 1953)
Part German, Irish, Swedish, one eighth Cherokee, and one sixteenth Jamaican

Angelina Jolie (b. 1973)
Czech and English descent on father's side and French Canadian and Iroquois on her mother's

EUROPEAN

Brigitte Bardot (b. 1934)
Real name: Camille Javal

Greta Garbo (1905–90)
Latherer in men's barbershop

Audrey Hepburn (1929–93)
Real name: Audrey Kathleen Ruston
Pseudonym used during World War II: Edda van Heemstra
Descendant of King Richard III of England (1312–1377)
Dancer with Ballet Rambert
Appeared in her first film as Edda Hepburn

Gina Lollobrigida (b. 1927)
As a fashion model she used the name Diane Loris
Lollo rosso lettuce is named after her 1960s hairstyle.

BRITISH

Michael Caine (b. 1933)
Real name: Maurice Micklewhite

Sean Connery (b. 1930)
Milkman
Winner of Mr. Scotland competition

Daniel Day-Lewis (b. 1957)
Son of the poet laureate Cecil Day-Lewis

Alec Guinness (1914–2000)
Real name: Alec Guinness de Cuffe
Advertising copywriter

Benny Hill (1924–92)
Milkman in the 1940s (before Ernie "the Fastest Milkman in the West," his most famous character)

Glenda Jackson (b. 1936)
Checkout till operator at Boots the Chemist

Sir Ben Kingsley (b. 1943)
Real name: Krishna Bhanji

Christopher Lee (b. 1922)
Served in the RAF—mentioned in dispatches
Served in British intelligence

James Mason (1909–84)
Architect

Peter Ustinov (1921–2004)
White Russian
Father was British spy

WRITERS

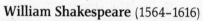

William Shakespeare (1564–1616)
Poacher on Charlecote Park, Sir Thomas Lucy's estate in Warwickshire.
In later years, Shakespeare was invited back to the estate on a deer shoot, but was so nearsighted that he was not allowed to handle a crossbow.

Edgar Allan Poe (1809–49)
Classics scholar

Daniel Defoe (1660–1731)
Reporter
Spy

Mark Twain (1835–1910)
Real name: Samuel Langhorne Clemens
He adopted Mark Twain as his pen name.
Gold miner
Reporter
Editor
Mississippi riverboat pilot
Twain had heard the leadsman on the paddle steamers calling out the depths of the river, from knots, or "marks," on a line hanging over the side. Mark "twain" indicated the second knot, which was two fathoms or twelve feet deep, the safe depth of water for a steamboat.

Barbara Taylor Bradford (b. 1933)
At age twenty became an editor on Fleet Street, London

SINGERS

Rod Stewart (b. 1945)
Grave digger

Stevie Wonder (b. 1950)
Real name: Steveland Judkins

Madonna (b. 1958)
Real name: Madonna Louise Ciccone

GREAT ARTISTS

Leonardo da Vinci (1452–1519)
Born illegitimate (illegitimacy was not a stigma during the
Renaissance)
His first commercial painting was to decorate a shield, for
which his father received 100 ducats (and kept it all).

Michelangelo (1475–1564)
Michelangelo di Ludovico Buonarroti Simoni

Raphael (1483–1520)
Raffaello Sanzio d'Urbino

Rembrandt (1606 or 1607–69)
Rembrandt Harmenszoon van Rijn

Caravaggio (1571–1610)
Michaelangelo Merisi
Double murderer

OTHERS

Mother Teresa (1910–97)
Real name: Agnes Gonxha Bojaxhiu (Albania)
    "The biggest disease today is not leprosy
        or tuberculosis, but rather the feeling
            of being unwanted."

Saint Matthew (first century)
Tax collector

Saint Mark (first century)
Born in Libya
Wrote the first of the Gospels in Greek

Albert Einstein (1879–1955)
Swiss Patent Office clerk (second class)
When he was asked to explain his theory of relativity,
Einstein said,

> "When you sit with a pretty girl for two hours,
> you think it's only a minute. When you sit
> on a hot stove for a minute, you think
> it's two hours. That's relativity."

Thomas Alva Edison (1847–1931)
Telegraph operator

Louis Vuitton (1821–92)
Opened first shop in Paris in 1854
Best customer: Empress Eugénie

Christine Keeler (b. 1942)
Nude dancer and hostess at Murray's Cabaret Club
> "Discretion is the polite word for hypocrisy."

Casanova (1725–98)
Trainee abbot
Spy for Louis XV of France
Founder of the French state lottery

Charles Manson (b. 1934)
Failed an audition to be a member of the Monkees pop
group (but this may be an urban myth)

FOOD AND DRINK

COVERING AGRICULTURE, COOKERY, TYPES OF FOOD,
DIETS, BEVERAGES, AND KITCHEN EQUIPMENT

AGRICULTURE

It is believed that agriculture—cultivating soil, harvesting crops, and raising livestock—began in the Middle East about 9000 to 7000 BC. The ox-drawn plow was invented in Mesopotamia (modern-day Iraq) around 4500 BC.

COOKERY

The origins of cookery are uncertain, but in all probability it began in the Paleolithic period. Cooked food is thought to have originated from the chance discovery of burned animal carcasses after a forest fire. The meat would have been tastier and easier to chew than it was when raw.

FRENCH CUISINE

For most of the twentieth century, French cuisine has been recognized as the most sophisticated in the West. Its preeminence dates to 1533 when the Florentine Catherine de Médicis (1519–89), at the age of fourteen, married the Duc d'Orléans, (1519–59), who was later to become Henry II of France. Catherine brought her staff of Florentine chefs, who taught the French a thing or two about refining their, up until then, rather crude methods of cooking.

COOKBOOKS

The first recorded cookbook was written by Archestratus of Gela, a Sicilian Greek who lived around 350 BC. He called his book *Hedypatheia* (*The Life of Luxury*).

Archestratus traveled throughout the Greek world of Sicily, Italy, Asia Minor, and Greece to record recipes. He emphasized the use of fresh seasonal ingredients with sauces to enhance flavors. He also recorded cooking techniques and combinations of flavors from around the empire.

In the second century BC, the Greek grammarian and author Athenaeus wrote the *Deipnosophistae* (*The Learned Banquet*). The fifteen-volume book contained many wonderful recipes, but was not strictly speaking a cookbook since it was written in a style more like a novel. In the book, Athenaeus imagines learned men, including some real people from the past, meeting at a banquet and discussing food and other subjects. The work covers most aspects of the ancient Greek and Roman world, and contains around eight hundred quotations from writers from antiquity.

The earliest and most important Latin cookbook *De Re Coquinaria* (*On Cookery*) was written by Marcus Gavius Apicius (14 BC–AD 37). In his book, Apicius showed the changes in taste and style of the Roman upper class leading up to the fall of the Roman Empire. Some of the dishes Apicius wrote about still feature in regional Italian food. Pliny the Younger (circa AD 62–113) claimed that Apicius force-fed geese to enlarge their livers to produce the best pâté, the forerunner of pâté de foie gras.

Recipes in *De Re Coquinaria* included a casserole of flamingo and nightingale tongues. Apicius entertained lavishly but lived beyond his means and found himself in financial difficulties. Sad to say, he decided to poison himself rather than face the consequences.

In the thirteenth century Kublai Khan's (1215–94) personal chef, Huou, wrote *The Important Things to Know About Eating and Drinking*. The majority of the book was dedicated to a collection of soup recipes, with household advice thrown in.

The first American cookbook was the delightfully titled *American Cookery, or the Art of Dressing* written by Amelia Simmons and first published in 1796.

TYPES OF FOOD

BREAD

Loaves have been found in 5,000-year-old Egyptian tombs, which are displayed in the British Museum, but the first evidence of bread goes back further still.

The origin of bread lies in the Neolithic Stone Age, with people who made solid cakes from stone-crushed barley and wheat about 12,000 years ago. Archaeologists have discovered a millstone, thought to be more than 7,000 years old, which was used for grinding grain.

 The Bible contains several references to bread, including the parable of the loaves and fishes, and the ancient Greeks and Romans used both unleavened and leavened (risen) bread as a staple food.

PIZZA

The basis of pizza, unleavened bread, has been around for centuries, and historical writings record that the ancient Egyptians, Greeks, and Romans all used forms of flat unleavened bread as a base for vegetables.

The invention of the first modern pizza is credited to Neapolitan restaurant owner Raffaele Esposito. His pizza Margherita combining pizza crust, tomato sauce, mozzarella cheese, and basil—the red, white, and green of which matched the colors of the Italian flag—was produced to commemorate the visit of Queen Margherita di Savoia (1851–1926) to Naples in 1889.

The first pizzeria in the United States was opened by Gennaro Lombardi on Spring Street, New York City, in 1905.

Chicago deep-dish pizza is jointly credited to Ike Sewell, a Texan businessman, and Ric Riccardo, who opened Pizzeria Uno, serving their newly invented pizza in 1943.

POTATOES

The potato was cultivated in South America for approximately two thousand years before it was introduced into Europe by the Spanish. Potatoes were brought to England from the West Indies by Sir Walter Raleigh (1552–1618).

TOMATOES

Cultivated in South America, tomatoes were introduced into Europe by the Spanish in the early part of the sixteenth century. Homegrown tomatoes were first eaten in England in 1596, from the garden of John Gerard, a barber-surgeon.

CHEESES

How can you govern a country that has
two hundred and forty-six varieties of cheese?
—CHARLES DE GAULLE (1890–1970)

The first cheese is thought to have developed around 4000 BC as a result of Sumerian herdsmen storing their daily ration of milk in dried calf stomachs. The milk combined with the natural enzyme rennin left in the stomach and then curdled, becoming cheese.

Cheese was described by Homer in the *Odyssey* in about the seventh century BC.

In Greece, whey was separated from curds by using a wicker basket called a *formos*, which is the root of *fromage*, the French word for cheese.

Stilton was first made in the mid-1700s in Melton Mowbray by Mrs. Frances Pawlett (1720–1808). It was Britain's first blue cheese and remains a market leader.

COOKIES (BISCUITS)

In the United Kingdom cookies are known as biscuits. The word *biscuit* derives from the Latin *bis coctum*, meaning "cooked twice."

The biscuit is first recorded in the twelfth century. Richard I (1157–99), also known as Richard the Lion-hearted) took "biskits of muslin" to the Crusades.

Beginning in the seventeenth century, the English navy supplied mass-produced biscuits for its sailors because they lasted longer than bread on extended sea voyages.

The Garibaldi biscuit was baked to commemorate the visit of Giuseppi Garibaldi (1807–82), founder of modern Italy, to London in 1864.

Cookies (biscuits) were first brought to America by British and Dutch immigrants in the seventeenth century.

The first known recipe for brownies was published in the Sears Roebuck catalogue of 1897, and the same catalogue sold the brownie mix.

Chocolate chip cookies were invented by Ruth Graves Wakefield of Whitman, Massachusetts, in 1937. She chopped pieces of sweet chocolate into cookie dough, assuming it would melt into the mixture during cooking. To Wakefield's surprise, the chocolate held its shape, a taste sensation was born, and the rest is history.

Fortune cookies are an American invention, introduced by the so-called forty-niners (Chinese laborers working in the 1849 California gold rush), and were not eaten in China until the 1990s when they were advertised as "Genuine American Fortune Cookies."

SAUSAGES

The word *sausage* is derived from the Latin *salcicius*, meaning "salted and preserved meat," not necessarily in a skin, which was to be cooked and eaten hot. Various forms of sausage were known in Babylonia, ancient Greece, and Rome.

Homer mentions goat meat sausages in the *Odyssey* in about 700 BC, and ancient Greeks were certainly eating cooked sausages by 9 BC. Sausages were introduced into Britain by the Romans.

Sausages were known as "little bags of surprises" in

Victorian England. This expression came from the uncertain contents. In the days when product labeling lay nearly a century in the future, it was not unheard of for sawdust to be included.

The origin of the nickname *banger* for a sausage came about during the Second World War, when sausages contained so much water that they had a tendency to explode during frying.

ICE CREAM

In the first century, the Roman emperor Nero (AD 37–68) ordered runners to pass buckets of snow from the mountains in the north, along the Appian Way, down to Rome. The snow was mixed with red wine and honey to be served at banquets.

The Chinese may have invented a form of half-frozen, fruit-flavored ice cream in the first millennium. Marco Polo (1254–1324) returned to Venice from his trip to the Far East with ancient recipes for concoctions made of snow, fruit juice, and fruit pulp.

The first documented record of milk being added to the icy slush to produce a form of modern ice cream was in 1672, when it was served to King Charles II (1630–85) of England.

The first company to sell ice cream from tricycles was Wall's in 1924. The new means of distribution was launched with the slogan "Stop me and buy one."

The Wall's company sold sausages in winter and ice cream in summer to "equalize the seasonality."

MARGARINE

Margarine was invented and patented in 1869 by the exotically named food chemist Hippolyte Mège-Mouriès (1817–80). The first mixture consisted of beef fat, cow's udder, and chopped sheep's stomach. The French government had offered a prize to anyone finding an acceptable substitute for butter, and the deeply unappetizing Mège-Mouriès creation won.

Modern margarine was created in 1915 by adding hydrogen to the mixture to harden it to the consistency of butter. The raw ingredients were almost irrelevant, as flavorings were added at a later stage of the process.

CANNED FOOD

In 1810, Nicholas François Appert (1750–1841), a French chef, published a method for preserving food in tin cans. With his invention, Appert won the 12,000 franc prize that Napoléon Bonaparte had offered for the best method of preserving food for his troops on long marches.

FROZEN FOOD

The practice of freezing food to preserve it can be traced back to 1626, and commercial production of frozen food began in 1875, but early food-freezing methods suffered by freezing the food too slowly. Slow freezing broke down the cell walls of the foodstuff and failed to preserve texture, appearance, and flavor.

In the 1920s, Clarence Birdseye (1886–1956) developed two methods for quick-freezing fish. While he was working as a fur trader in Canada, Birdseye had observed how the Inuit people preserved fish by rapid freezing, in readiness for the harsh Arctic winters. He bought a simple electric fan and, using buckets of salt water and ice, devized an industrial method of flash-freezing food under pressure and packing it in waxed cardboard boxes. In 1924, Birdseye put his first frozen fish on sale and founded the frozen food industry as it has evolved today.

Starting with an initial investment of only $7, Birdseye sold out his patents in 1929 for $22 million. The product range was rebranded Birds Eye.

CHOCOLATE

The Aztecs and Mayans discovered the stimulation value of the cacao tree in around AD 600 and made a nourishing drink from the cocoa beans. They called it *xcoatl*, and in 1519 served the bitter drink to Hernán Cortés (1485–1547), the Spanish conquistador. He took it with him on his return to Spain, where it remained secret for more than one hundred years.

The first chocolate factory in America was set up by John Hannan and James Baker in Dorchester, Massachusetts, in 1765.

The first chocolate bar in Britain was sold by Fry & Sons in 1847. John Cadbury (1801–89) started selling chocolate in 1849 at Bingley Hall in Birmingham.

CHEWING GUM

The ancient Greeks chewed mastic gum, a product of the mastic tree. Other ancient cultures in India and South America also enjoyed the benefits of chewing raw gum.

The first modern chewing gum was invented by Thomas Adams of New York in 1869 after a meeting with Antonio Lopez de Santa Anna (1794–1876), the exiled ex-president of Mexico, who told him of chicle gum, which the native Mexicans had been chewing for years. Initially, Adams tried to market a blend of chicle gum and rubber as a substitute material for carriage tires, but this failed. He then began selling it as a flavorless but chewy ball to be used as chewing gum, calling his product Chiclets.

POTATO CHIPS

American restaurant owner George Crum invented potato chips in 1853 after the railway and shipping magnate Cornelius Vanderbilt (1794–1877) had complained about the thickness of the french fries being served in Crum's Saratoga Springs, New York, restaurant. Crum was angered by these remarks and decided to serve up deep-fried, paper-thin potato chips. To Crum's surprise, Vanderbilt approved the new product. Crum's potato chips became a well-known delicacy, which he called Saratoga chips.

Mass-marketing of potato chips began in the United States in 1926 when Laura Scudder began to sell them in waxed paper bags. Scudder invented the airtight bag to keep the chips fresh, by ironing together pieces of waxed paper. Until that time, potato chips were distributed in large tins, and the last chips out of the tin were usually stale.

DIETS

The earliest recorded "fad" diet was that followed by William the Conqueror (1028–87). William grew so fat that he confined himself to his bedroom, ate no food, and drank only alcoholic drinks. In 1087, at the

Battle of Mantes, near Rouen in France, the strap holding the saddle on his horse gave way under the strain, and William died from the injuries he suffered as he fell on the pommel of the saddle.

William Banting (1796–1878) is the first person known to have deliberately regulated his weight by controlling his food intake. Banting's diet, which included reducing his carbohydrate consumption, was supervised by Dr. William Harvey. The results were first published in Banting's booklet, *Letter on Corpulence Addressed to the Public*, in 1863. The booklet went on to become a worldwide bestseller.

The Atkins diet, which achieved cult status in 2001–2, began with the publication of *Dr. Atkins' Diet Revolution* in 1972 by Dr. Robert Atkins (1930–2003). At his death, Atkins weighed 260 pounds.

Weight Watchers began in 1961 with the efforts to lose weight of an obese Brooklyn housewife, Jean Nidetch (b. 1923). Nidetch held meetings in her home to discuss ways of losing weight, and within a few months people were lining up on the street to attend. In 1963, Nidetch created Weight Watchers, and formalized the ideas that had developed from her meetings.

BEVERAGES

TEA

According to ancient myth, in 2737 BC, a handful of dried leaves from a tea bush blew into a pot of boiling water, into which the Chinese emperor Shen Nung (Divine Farmer) was staring (see "Health," p. 125). There

is no record to say why the pot was being boiled or why the emperor was staring at it, nor why the leaves had been dried. The story relates that the resulting brew was henceforth known as *tchai* and became China's national drink during the T'ang dynasty (AD 618–907).

Outside China, the Arabs were the first to mention tea, in AD 850.

The first time tea was seen in Europe was in 1560, when it was introduced into Portugal by Father Jasper de Cruz, a Jesuit priest.

Tea was first sold publicly in England in 1657. Milk was not added to the brew until 1680.

The first tea imports to America were in 1650 by Peter Stuyvesant (1612–72), the last Dutch director general of New Amsterdam (New York).

The most famous tea party was the so-called Boston Tea Party of 1773, in which a cargo of tea was emptied into Boston Harbor in protest against excessive English excise duty. The protest is best known for signaling the start of the American War of Independence (see "War," p. 271).

Iced tea was first served in 1904 at the St. Louis World's Fair, by a tea plantation owner.

Tea bags were accidentally invented by American tea importer Thomas Sullivan, who gave his customers samples of tea in silk bags. Mistakenly, they put the entire bags into the pot and the tea bag was born.

Thomas (later Sir) J. Lipton (1850–1931) patented the tea bag in 1903 and a four-sided tea bag in 1952.

COFFEE

Before AD 1000, the Galla, or Dromo, of Ethiopia began to eat ground coffee beans mixed with animal fat for extra energy. It is recorded that a goatherd called Kaldi noticed his goats jumping around with increased energy after chewing berries from the coffee bushes growing wild in his fields. He tried them and found renewed vigor when he became tired, supposedly becoming the first person to benefit from a shot of caffeine. News of the added energy to be gained from Kaldi's "magic" beans spread rapidly throughout the region, and coffee consumption became a national habit.

Coffee was first imported to Constantinople in 1453 by the Ottoman Turks, and the world's first coffee shop, Kiva Han, opened there in 1475. By the sixteenth century, coffee drinking had spread throughout Asia, and coffeehouses began to serve as public gathering places in Persia, in the same way as taverns.

Edward Lloyd opened his coffeehouse in London in 1668. It was used principally by merchants, seafarers, and bankers and eventually grew to become Lloyd's of London, the center of the world insurance market.

Ninety percent of the world's coffee production can be traced back to a single bush, stolen by French naval officer Gabriel Mathieu de Clieu in 1723. The seedling was transported to Martinique in the Caribbean, and within fifty years, almost 20 million bushes had been propagated.

The first coffee drunk in America was in 1607, introduced by Captain John Smith, one of the founders of Virginia. Later, after the Boston Tea Party in 1773, coffee drinking became a patriotic duty, so that British tea would fall out of favor and lose its market.

The Maxwell House brand was founded in 1886. It was named in honor of the Maxwell House Hotel in downtown Nashville. The coffee served there was blended by Joel Cheek, a local wholesale grocer. The hotel burned down in 1961.

The first instant coffee was produced in Chicago in 1901 by chemist Satori Kato.

The first decaffeinated coffee was marketed in Germany and France under the Sanka (a contraction of *sans caffeine*) brand in 1906. The product was the result of experiments organized by food scientist Ludwig Roselius on a batch of ruined coffee beans. Roselius, one of the founders of the Kraft Food Company, introduced Sanka into the United States in 1923.

The first freeze-dried coffee (under the Nescafé brand) was introduced in 1923 in Switzerland. Nescafé supposedly called its Blend 37 brand after Didier Cambreson, who completed the whole twenty-four hours of the 1937 Le Mans race on his own after his codriver failed to show up. Cambreson was driving Car 37 and came in thirty-seventh place. It is claimed he kept going by drinking only the new Nescafé brand.

Starbucks opened its first coffee shop in 1971 in Pike Place Market, Seattle. The original coffee shop named after the first mate in *Moby Dick*, was started by an English teacher, Jerry Baldwin; a history teacher, Zev Siegel; and a writer, Gordon Bowker, and is still trading. By 1992, when it floated as a public company, it had 165 coffee shops. In 2005, the total stood at more than 11,000 outlets worldwide.

BEER

Tests on ancient pottery jars from present-day Iran reveal that they were in use for holding beer up to seven thousand years ago, and archaeologists have unearthed Babylonian clay tablets dated 4300 BC, revealing the earliest recipe for producing beer.

The early Romans were drinking beer from the fifth century BC.

During the late Middle Ages in England, beer was used as a form of payment, even of taxes.

The first beer brewed in America was on the estate of Sir Walter Raleigh in 1587. The colonists had sent messages to England requesting better beer supplies, but then decided to brew their own.

The first commercial brewery in America was set up in New Amsterdam (now New York) in 1612.

There were more than two thousand breweries in the United States in 1880. By 1935 the number had shrunk to 160. By 1992 most of the breweries had been consolidated into just five giants that controlled 90 percent of U.S. beer consumption.

Lagering techniques were developed in the period of 1820–40, resulting in a lighter tasting beer, but still with dark coloring from hard German water. The first modern lager beer, with a characteristically light golden color, was produced in Pilsen, Bohemia (now in the Czech Republic), in 1842, by twenty-nine-year-old brewer Joseph Groll. Groll used the soft local water to produce his lager. The type of beer was named after the town, and became pilsener or pilsner.

The first canned beer was Kreuger Cream Ale, sold in 1935 by the Kreuger Brewing Company of Richmond, Virginia.

Guinness, which is a stout, has been brewed at the St. James's Brewery in Dublin since 1759. Arthur Guinness signed a nine-thousand-year lease for forty-five pounds a year. Contrary to the myth that the Guinness brewed in Dublin uses water from the river Liffey, it actually comes exclusively from the Lady's Well in the Wicklow Mountains.

Beer widgets

Brewers struggled for years to devise a way to produce bottled and canned beer, which, when poured, would successfully imitate the creamy head on draught beer. The invention of the beer widget solved the problem in 1985.

The widget was invented by William Byrne and Alan Forage, and patented by the Irish brewer Guinness. It works by releasing nitrogen into the beer to produce foam after the can or bottle has been opened.

Draught Guinness in a can, the first beer with a widget, was launched in March 1989.

The original use of the term *widget* was in the 1924 play *Buxton on Horseback*, as a product manufactured by one of the characters.

WINE

The Neolithic people of the Near East and Egypt are credited with producing the first wine between 8500 and 4000 BC. Pottery jars used for storing wine were developed around 6000 BC.

Buried jars, dated between 6000 and 7000 BC, have been unearthed in the Chinese village of Jiahu, containing a winelike liquid.

Winemaking is recorded on the walls of Egyptian tombs dating from 2700 BC and shown to be part of ceremonial life. Since grapes were not grown in ancient Egypt, it is most likely they were imported from the Near East.

Champagne

> *I drink champagne when I win, to celebrate, and drink*
> *champagne when I lose, to console myself.*
> —NAPOLÉON BONAPARTE (1769–1821)

The creation of the sparkling wine champagne was supposedly a collaborative effort between two great cellar masters of the monastic orders of Pierry and Epernay in the Champagne area of northeast France.

With their abbeys only two miles apart, Frère Jean Oudart and Dom Pierre Pérignon (1638–1715) are jointly credited with producing the first successful champagne in the late seventeenth century. Although the principles established by Dom Pérignon hold good today, modern dating techniques suggest the earliest sparkling wine was produced around 1535.

SCOTCH WHISKEY

The origins of Scotch are uncertain, although it is known that the ancient Celts in Ireland understood distilling before AD 1200. They called their drink *uisge beatha* (the water of life).

The first evidence of a distilling process, from the eighth century BC, has been uncovered in China.

By the sixth century, distilling was taking place in England, but the first mention of distilling in Scotland appears in the exchequer rolls of 1494, which list sufficient malt to produce 1,500 bottles of Scotch. This suggests that an industry was already supplying an established demand.

BRANDY

Although there is some evidence of brandy being known in the twelfth century, it is generally thought that an unknown Dutch trader invented brandy in the sixteenth century. In an effort to save storage space, he started boiling his wine to cause the water in it to evaporate. The resultant liquor became *brandewijn* meaning "burnt wine."

KITCHEN EQUIPMENT

REFRIGERATOR

The freezing properties of ammonia were discovered in 1868 by French scientist Charles Tellier (1828–1913), leading directly to the development of the refrigerator. Unfortunately for him, he failed to patent his idea and ended up dying in poverty. (See "Inventions," p. 156).

TOASTER

Making toast in front of an open fire began as a way of using stale bread and was known to the ancient Romans.

The electric toaster was invented in 1891 in England and produced by Crompton & Co.

MICROWAVE OVEN

In early 1946, Dr. Percy Spencer (1894–1970) of Raytheon Corporation accidentally discovered the possibility of cooking by microwave during experimental work with magnetrons, which produced microwaves for use in radar. He found chocolate bars in his jacket pocket had mysteriously melted when he ran low-level microwaves.

The patent for the microwave oven was filed in late 1946, and the first oven, branded the Radarange, went on to the professional catering market in 1947. The Radarange stood over six feet tall and sold for more than $5,000.

Tappan Corporation introduced the first domestic microwave ovens in 1955 at $1,295 each.

HEALTH

COVERING HISTORY OF MEDICINE, DISEASES AND CURES, EARS, EYES, HEART, KIDNEYS, MIND, TEETH, AND DRUGS

When you don't have any money, the problem's food.
When you have money, it's sex.
When you have both, it's health.
—J. P. DONLEAVY (B. 1926)

Health is one of our biggest preoccupations. This chapter highlights the origins of the main developments in health and the battle against diseases, with thumbnail sketches of the routes some pioneers took.

HISTORY OF MEDICINE

One cannot practice a science well
unless one knows its history.
—AUGUSTE COMTE (1798-1857)

PREHISTORY

Neolithic man used stone tools with a cutting edge to lance abscesses and let blood. Evidence has also been found in the Petit-Morin valley in France that around 20,000 BC, using the same stone tools, Neolithic man was also performing delicate and successful operations, such as removing disks of bone from skulls. The holes in the skulls show evidence of healing, which indicates that the patient survived. Whether this very difficult operation was for the purpose of releasing evil spirits or for the relief of pressure inside the skull is unknown, although it is thought that magic formed a major part of medical treatments, and that the doctor and the priest would have been the same person.

Six thousand years ago at Ur in Mesopotamia, the ancient Sumerians had begun to base their medicine on astrology.

CHINESE

Traditionally, the Chinese have tested the effectiveness of herbal remedies on themselves rather than using animals. The earliest known practitioner of herbal medicine was the emperor Shen Nung, who lived around 2800 BC, and who is reputed to have tasted hundreds of different herbs in his quest for medical cures. Shen Nung is venerated as the father of Chinese medicine, and is thought by some scholars to have invented acupuncture.

The system of yin and yang, the belief that nature is in harmony when the yin and yang (female and male) elements are in balance, is also attributed to Shen Nung.

Chinese legend also records that Shen Nung invented the plow and was the first to drink tea after some dried tea leaves blew into his cup of boiling water one day (see "Food and Drink," p. 117).

BABYLONIAN AND EGYPTIAN

By 2000 BC the ancient Babylonians had begun to base their medicine on religion. Before that surgeons were answerable directly to the state, and the code of the profession had established savage punishments for medical failure, whereas when practicing medicine as priests the surgeons were answerable only to God.

In 1500 BC, the ancient Egyptians began the practice of cauterizing—sealing by burning—wounds, to stop bleeding.

INDIAN

Ayurveda, the ancient Indian system of healing, is the oldest holistic (whole body) system in existence. It was first expounded by Dhanvantari around 1500 BC, but not written down until 200 BC. The Ayurvedic system teaches the elements of surgery as well as knowledge of plants, herbs, aromas, colors, and lifestyle. It is thought that Ayurveda may have developed out of earlier texts, dating from 3000 BC, which included instructions in spirituality and behavior. The *Atreya Samhita*, one of the Ayurvedic books, is the oldest medical book in the world.

Injuries to the nose were common in the first millennium BC. The maiming of prisoners of war by cutting off their noses, and the slitting of noses in combat was common. In 500 BC Sushruta of India was the first to conduct a successful rhinoplasty (reshaping the human nose). Sushruta described how a strip of skin could be left partly connected to the forehead and grown over the place where the nose had been cut off.

GREEK

Ancient Greece consisted of a collection of city-states, which were independent of one another. Despite this independence, the states shared common ground in their attitudes to healthy lifestyles and the treatment of illness.

Aesculapius According to ancient Greek mythology, the greatest of the ancient Greek surgeons was Aesculapius, who was born a mortal but made into a god in the fifth century BC. Aesculapius was said to be the son of Apollo, who himself was regarded as the god of medicine until Aesculapius usurped that role.

Alcmaeon of Croton The dawn of scientific medicine came to Greece between 600 and 700 BC. In 550 BC, Alcmaeon of Croton (b. 535 BC) pioneered experimental medicine and produced the first ear trumpet. He also discovered the optic nerve and was the first to consider the brain to be the center of intellectual activity.

Pythagoras In 530 BC, the Greek philosopher and mathematician Pythagoras (582–496 BC) founded the world's first medical school in the town of Croton, in what is present-day southern Italy. The school was founded around the time when the practice of medicine had begun to acquire professional status.

Hippocrates The Greek physician Hippocrates (460–380 BC), who was born and brought up on the island of Kos, wholly rejected the superstition and magic of primitive medicine. Hippocrates was the first to introduce scientific method to the treatment of illnesses and recommended that doctors should accurately record all their treatments for the benefit of other doctors who followed. Hippocrates also introduced the idea of patient confidentiality and is widely regarded as the father of medicine.

Doctors still swear the Hippocratic oath, which is regarded as the foundation of Western medical ethics: "I swear by Apollo the healer, by Aesculapius, by health and all the powers of healing, that I will use my power to help the sick to the best of my ability and judgment."

Herophilus of Chalcedon The first public dissection of a human body was performed by Herophilus of Chalcedon (320–260 BC) in 300 BC. He is credited with being the first to carry out scientific explorations of what lay within the human body and is regarded as the father of human anatomy. He was the first to challenge Aristotle's theory and consider the brain, not the heart, as the seat of consciousness, and was the first to distinguish between motor and sensory nerves. The study of anatomy advanced enormously under his guidance and was continued by his rival Eristratus. With their deaths, advances in human anatomy ground to a halt until Leonardo da Vinci (1452–1519) accurately drew the inner workings of the body, more than a thousand years later.

Galen of Pergamum The greatest Greek contributions to the furthering of medical knowledge came from Galen of Pergamum (129–200 BC). He wrote more than four hundred treatises, of which three hundred were lost in a fire. He proposed that careful hygiene, good diet, and plenty of exercise were important factors in health care, and made an outstanding contribution to the understanding of the cardiovascular system. Galen was summoned to become surgeon to the Roman emperor Marcus Aurelius (AD 121–80), and after Marcus's death, to his son, the emperor Commodus (AD 161–92).

Soranus of Ephesus Known as the birthing doctor, between AD 98 and 138 Soranus of Ephesus was the first to make an in-depth study of obstetrics, the branch of medicine dealing with the care and treatment of women before, during, and after childbirth. Soranus realized that women's reproductive anatomy led to unique medical problems that men could not experience. He fought against superstition and misunderstandings for most of his life, writing widely on the delivery of babies.

ROMAN

In AD 379, after a severe famine, Saint Basil the Great (AD 329–79), who had been a school friend of the Roman emperor Julian (AD 331–63), sold his family land, bought food to feed the starving, and founded the world's first hospital in Caesarea, in what is present-day Israel.

The first European hospital was opened in Rome around AD 400. Funding for the hospital was provided by Fabiola, later Saint Fabiola (d. AD 399), a Roman noblewoman who was a member of the wealthy Fabia family. Divorced from her first husband, who had abused her, and left widowed by the death of her second husband, Fabiola had ambitions to become a hermit and live in Jerusalem. She traveled to Bethlehem in AD 395, but her involvement with the hospital led her to return and she became a nurse instead.

GENERAL

The first English hospital was built in York in AD 936 and was originally dedicated to Saint Peter. The hospital was rededicated to Saint Leonard in 1137 after a fire, but it closed four hundred years later with the dissolution of the monasteries.

The first European medical school was established at Montpellier in southern France in 1220. In 1181, the lord of Montpellier had given his permission for anyone to come and teach medicine within Montpellier, no matter where they came from. The school attracted teachers and students from as far away as Scotland.

The first organized school of medical teaching was started in about 1260 by Thaddeus of Florence (1223–1303), who taught at the University of Bologna.

The first hospital in the New World was founded in Mexico in 1524 by Hernán Cortés (1485–1547), the Spanish conqueror.

The first animal-to-human bone graft was performed in 1668 by Job van Meekeren of the Netherlands when he grafted part of a dog's skull into the wounded leg of a Russian soldier.

Plaster of paris was first used as a bandage material in 1852 by Antonius Mathijsen of the Netherlands.

The speech center of the human brain was discovered in 1861 by the French doctor Paul Broca (1824–80).

The first to use plaster of paris to treat spinal injuries was American orthopedic surgeon Lewis Sayre (1820–1900), in 1877.

The first appendix operation in Britain was carried out by Dr. H. Hancock in 1848, although there are reports of a British Army surgeon called Amyan performing an appendectomy as early as 1735.

The first successful appendix operation in the United States was carried out in Davenport, Iowa, in 1885 by Dr. William West Grant. The patient was twenty-two-year-old Mary Gartside.

The first artificial incubator for premature babies was developed in 1888 by German obstetrician Karl Crede (1819–92). Electricity was not widely available, and the air inside had to be warmed by a kerosene lamp.

The first lumbar puncture operations—driving a needle into a patient's spine to obtain sample spinal fluid—were conducted independently in 1891 by Walter Wynter in England and Heinrich Quincke in Germany. Anesthetics were not available.

The first surgeon to use rubber gloves while performing an operation was William Halstead of Johns Hopkins Hospital in the United States, in 1894.

The surgical mask was invented by Englishman William Hunter in 1900.

The World Health Organization was established in 1948.

X-RAYS

While he was investigating the properties of light in his laboratory at the University of Würzburg, in Germany, on November 8, 1895, Wilhelm

Conrad Roentgen (1845–1923) discovered X-rays. After seven weeks of further research, to satisfy himself that the results were both genuine and repeatable, Roentgen presented his report to the medical establishment. By January 1896, the world was gripped by "X-ray mania," and within a few months X-rays were being widely used as diagnostic tools. Acclaimed as a miracle worker, Roentgen refused to patent his discovery, and in 1901 he was awarded the first Nobel Prize in Physics.

The first ever X-ray picture was of Roentgen's wife's left hand. The image clearly shows her wedding ring.

DNA

More properly known as deoxyribonucleic acid, DNA was made famous the world over in 1953 when James Watson (b. 1928), Francis Crick (1916–2004), and Maurice Wilkins (1916–2004) announced they had discovered its structure.

Watson, Crick, and Wilkins received the Nobel Prize in Physiology or Medicine for their discovery of the famous double helix shape of the DNA molecule. However, they most definitely did not discover DNA itself. It had actually been identified in 1869 by Swiss scientist Johann Friedrich Miescher (1844–95) working in Tübingen, Germany (see "Questionable Origins," p. 190).

CAT SCAN

The CAT (computerized axial tomography) scanner is a machine that takes X-rays from different angles. It was developed in the United Kingdom by EMI and launched in 1973. The price of individual machines started at £100,000. Godfrey Hounsfield (1919–2004), who led the development team, received the Nobel Prize in Physiology or Medicine in 1979 and was knighted in 1981.

IN VITRO FERTILIZATION

The world's first test-tube baby, Louise Joy Brown, was born in Oldham, Lancashire, on July 25, 1978. Louise's birth was made possible by the pioneering in vitro fertilization process. Elizabeth Carr, who was born in 1981, was the first American test-tube baby.

BLOOD

In 1628, William Harvey (1578–1657) published his most important book, *An Anatomical Study of the Motion of the Heart and of the Blood in Animals.* In the book, Harvey describes how blood is pumped around the body by the heart, returns to the heart, and is recirculated. The book was controversial at the time and lost Harvey many patients.

Blood transfusions were attempted in Europe in 1628 by Giovanni Colle, a professor in Padua, Italy, but because of the incompatibility of blood types it was not until 1654 that another Italian, Francesco Folli (1624–85), actually achieved the first successful blood transfusion.

It was not until the early twentieth century that the Viennese physician Karl Landsteiner (1868–1943) identified the four human blood types, enabling routine and non-life-threatening blood transfusions.

Blood pressure was discovered and reported in the preface to a book on plants written in 1727 by the Reverend Stephen Hales. Pressure was measured by inserting a tube directly into the vein of an animal and noting how high the blood rose in the tube.

Human blood pressure was first accurately measured in 1856 by J. Faivre of France, following the work of the German physician Karl von Vierordt (1818–85), who in 1855 developed the inflatable cuff to stop arterial blood flow so that blood pressure could be measured in a noninvasive way.

BRAIN

Early man has practiced brain surgery since Neolithic times, and Hippocrates himself left copious notes on how to treat head injuries and depression.

In 1926, the pioneering brain surgeon Harvey Cushing (1869–1939) was the first to use electrodes to cauterize blood vessels during brain surgery.

DISEASES AND CURES

SMALLPOX

An acute viral infectious disease, smallpox is characterized by fever and pockmarks on the skin, and could be fatal. It has now been eradicated.

The first vaccination for smallpox was developed in 1796 by Edward Jenner (1749–1823), and by 1801, 100,000 people had been vaccinated in England.

At the age of eight Jenner had survived an extremely risky treatment, which was supposed to prevent him from contracting smallpox for the rest of his life. He was subjected to variolation, the forerunner of inoculation, which had been introduced to Britain by Lady Mary Wortley Montagu (1689–1762), a noted eighteenth-century writer. The treatment involved taking pus from the sores in the body of a dead victim of smallpox and inoculating it directly into the patient. Variolation had been successfully used in China for centuries, but dosage was very difficult to control because the individual patient's reaction was impossible to predict. The treatment included three weeks' isolation from noninfected people.

After an intensive worldwide inoculation program conducted by the World Health Organization, it was announced on May 8, 1980, that smallpox, which had blighted humanity for thousands of years, had finally been eradicated from the world; the first, and so far only, disease ever to be conquered.

PASTEURIZATION

The process known now as pasteurization was invented in 1862 by French chemist Louis Pasteur (1822–95) after he had performed experiments proving his germ theory, which maintains that microorganizms cause fermentation and the formation of mold. After establishing the cause, Pasteur set about finding a means of preventing the formation of molds in liquids, such as milk, by pasteurization.

Pasteur also managed to save the French silkworm industry, which had been blighted for years by a disease called pebrine that killed great numbers of silkworms. By eliminating the microbe that caused the disease, Pasteur brought the pebrine under control.

TUBERCULOSIS

Hippocrates (460–380 BC), writing in about 410 BC, identified a form of tuberculosis called phthisis, and by the nineteenth century almost 25 percent of all deaths in Europe were being caused by the disease. In adults, the bacillus spreads in the lungs, destroys the respiratory tissues, and then attacks the air passages, at which stage the patient becomes infectious. In 1882, the German doctor Robert Koch (1843–1910) discovered the microbe that causes tuberculosis, but it was not until the development of the antibiotic streptomycin, in 1943, that successful treatment was possible.

Streptomycin was discovered in 1943 by Albert Schatz (1920–2005) after performing research at Rutgers University in New Jersey. His supervisor, Selman Waksman (1888–1973), took all the credit for the discovery, as well as the Nobel Prize in Physiology or Medicine in 1952. However, Schatz successfully sued Waksman for a share of the streptomycin royalties.

POLIO

A viral infection, polio attacks the muscle-controlling nerves of the brain and spinal cord, resulting in paralysis. There is no cure for the disease, but in 1952, Dr. Jonas Salk (1914–95) discovered a vaccine that would prevent its onset. The vaccine was made available to the public on April 12, 1955, and in a remarkably humane gesture Salk refused to patent his vaccine, holding that he had no desire to profit personally from his discovery, merely to see the widest possible distribution for the greatest possible good.

MALARIA

In 1880, the French physician Charles-Louis Alphonse Laveran (1845–1922) discovered that malaria is caused by a protozoan (a single-cell organism). For his work in the field he was awarded the Nobel Prize in Physiology or Medicine in 1907.

In 1898, Sir Ronald Ross (1857–1932) discovered the life cycle of the malaria parasite as it develops in the malaria mosquito and in the human host. He was awarded the Nobel Prize in Physiology or Medicine in 1902.

From the end of the Second World War to the 1970s, the pesticide DDT was used to destroy local mosquito populations, helping to eliminate malaria from certain parts of the world. However, a ban on DDT has been in force since then, and the mosquito population has steadily increased.

ACQUIRED IMMUNE DEFICIENCY SYNDROME (AIDS)

AIDS, originally called GRID (gay related immune deficiency), is a condition in which the body's immune system becomes weakened due to the contraction of the human immunodeficiency virus (HIV). Those infected are susceptible to infections that eventually result in death. It is now believed that HIV is a new virus to emerge in the human population. Previously, it was an ancestral virus that infected monkeys.

AIDS was officially recognized for the first time on June 18, 1981, in California and New York. It was originally thought to be a sexually transmitted disease restricted to homosexual men, but it was quickly discovered that it could be transmitted to anyone through transfusion of contaminated blood or the use of contaminated intravenous needles. AIDS can also be transmitted from mother to child during pregnancy and breast-feeding.

The earliest evidence of HIV infection is in a 1959 plasma sample taken from an adult male in the Democratic Republic of Congo.

There was false evidence that a twenty-five-year-old British sailor, David Carr, had died from AIDS in 1959. This was dismissed after exhaustive testing of preserved tissue samples on both sides of the Atlantic.

It is thought that the first American male to die of AIDS was a fifteen-year-old African American who died in 1969 in St. Louis, Missouri. Preserved tissue samples were shown to be HIV positive.

The first accidental victim of AIDS from blood tranfusion was the unfortunate Californian Don Coffee, in 1981.

EARS

HEARING AIDS

The first reference in literature to a hearing aid is a reference by Homer to a speaking trumpet in the *Iliad*.

The first man-made auditory tube or ear trumpet was made in 550 BC by the early Greek medical writer and scientist Alcmaeon of Cro-ton. These first examples were intended not to provide help to the hard of hearing, but as an aid to hearing at distance, such as at sea in times of war, or on the hunting field.

By approximately 300 BC the ancient Greeks were importing seashells into Phoenicia to be used as ear trumpets. The shells were hardened and then painted, to make them more marketable.

The first modern ear trumpet was described by the Belgian scientist and high-school rector Jean Leurechon (1591–1670) in his book *Récréations Mathématiques*, published in 1624.

The first British maker of ear trumpets was Bevan of London in 1715.

Around 1790 Alessandro Volta, who developed the electric battery, discovered that the auditory system could be stimulated by electricity, when he put metal rods into his own ears and gave himself a 50-volt electric shock. The result of Volta's experiment was that he heard noises like "a thick boiling soup."

Modern hearing aids owe their existence to the first commercial hearing aid to use a carbon microphone. It was produced by the Dictograph Company of the United States in 1898.

By 1954, hearing aids were small enough to be built into spectacle frames, and fully digital hearing aids, worn within the ear, were introduced in 1995.

The first cochlear implant was carried out on Rod Saunders in Melbourne, Australia. Professor Graeme Clark of Melbourne pioneered the technique during the 1970s.

The cochlear implant is often referred to as a bionic ear since it is effectively an implanted hearing aid for the profoundly deaf and the very hard of hearing. Unlike traditional hearing aids, the cochlear implant does not amplify sound, but stimulates the auditory nerves with minute electrical impulses.

The first cochlear implant in the United States was performed in 1984, after approval by the U.S. Federal Drug Administration.

EYES

Alhazen (Abu 'Ali al-Hasan ibn al-Haytham; AD 965–1039) was a brilliant Arab mathematician who was born and lived in Basra, Persia (now in present-day Iraq). Alhazen was the first to establish, through experimentation, that people see things because rays of light pass from an object to the eye. Until Alhazen's research, it had been generally thought that the eyes sent out invisible rays to detect objects. Alhazen also described the magnifying properties of lenses, but his observations were not appreciated until the thirteenth century through the work of the British scientist-monk Roger Bacon (1214–94).

GLASSES

The Roman emperor Nero (AD 37–68) used to look through a highly polished emerald to watch gladiators fighting. Polishing it into a lens shape gave the emerald magnifying properties.

The invention of glasses is attributed to Salvino d'Armato (d. 1317), who introduced his invention in Florence in 1268. His glasses had no side arms and had to be balanced or held on the nose.

The side arms on glasses were not added until the eighteenth century in Paris when short arms were added to a pair to hold them to the sides of the head. In 1727, the English optician Edward Scarlett extended the arms to fit over the ears.

The first painting to portray a person wearing glasses was by Thomasso de Modena in 1352, showing an elder of the church, Hugh de Provence, peering at a manuscript.

The first printed book on ophthalmology was published in 1474. The title of the book was *De Oculis Eorumque Egritudinibus et Curis*, which had been written by Benvenuto Grassi in the twelfth century.

The invention of the printing press by Johannes Gutenberg (1398–1468) in 1456 triggered the production of reading glasses as all ranks of society began to read the printed word.

Bifocals are popularly believed to have been invented in 1784 by the great American scientist and statesman Benjamin Franklin (1706–90). However, there is historical evidence that Samuel Pierce, an English optician, may have invented them in 1775.

The first U.S. president to wear glasses was George Washington (1732–99), who wore them from 1776. Washington was also noted for wearing false teeth made of wood.

Progressive lenses were introduced in the 1960s. They are bifocals in which the boundary between the reading and distance parts of the lens is graduated and invisible.

OPTICAL SURGERY

The removal of cataracts from the eye was pioneered in 1748 by Jacques Daviel (1693–1762), optician to Louis XV of France. He cut out the opaque lens from beneath the cornea, the hard coating of the eyeball. No anesthetics were used.

Spanish ophthalmologist Ignacio Barraquer Barraquer (1884–1965) devised a method of cataract removal by suction in 1917.

CONTACT LENSES

Leonardo da Vinci drew sketches in 1508 showing several forms of contact lens.

Many attempts were made to produce contact lenses, including by French philosopher René Descartes (1596–1650), but the first commercially available contact lenses were introduced by William Feinbloom, a New York optometrist, in 1936. Bifocal contact lenses became available in 1982.

HEART

In 1707, John Floyer (1649–1734), one of the three great medical pioneers of Staffordshire, England (the others were Erasmus Darwin and William Withering), invented a stopwatch to measure the human pulse.

PACEMAKER

The artificial heart pacemaker is designed to regulate the beating of the heart when the organ's natural pacemaker is not effective. In 1862, English surgeon W. H. Walshe suggested the use of electrical impulses to control the heart's rhythm. It fell to Canadian electrical engineer John Hopps to design and build the first pacemaker in 1950. The early models were fitted outside the body and had to be plugged into a wall socket.

The first pacemaker implanted into the body was designed and fitted by Dr. Rune Elmqvist (1906–96) at the Karolinska Hospital in Solna, Sweden, in 1958. The patient, Arne Larsson, survived until 2001, having been fitted with no fewer than twenty-two throughout his life.

HEART SURGERY

During the Second World War, doctors made enormous advances in blood transfusion, anesthetics, and antibiotics, which led to the development of modern surgery. A young U.S. Army surgeon, Dr. Dwight Harken, removed fragments of shrapnel from the still-beating hearts of soldiers by inserting his finger into the wound, locating the shrapnel, and pulling it out through the same hole.

The problem faced in open-heart surgery is that once the heart is stopped, to perform the surgery, the surgeon has only four minutes to complete the procedure before the patient's brain is irreparably damaged due to lack of oxygenated blood pumped by the heart. In 1952, Dr. Bill Bigelow (1913–2005) of the University of Minnesota, who had studied the habits of hibernating animals, had the idea of reducing the patient's temperature from ninety-eight to eighty-one degrees Fahrenheit. As a result, doctors were able to extend the operating time from four to ten minutes.

The heart-lung machine was developed by John Gibbon (1903–73) of

Philadelphia in 1953 to overcome these difficulties. By connecting the patient to the machine, the heart could remain in a still condition for the operation, allowing the surgeon ample time to perform the complex procedures.

The first successful open-heart surgery took place on May 15, 1953, on eighteen-year-old Cecilia Bavolek, who was connected to the heart-lung machine for twenty-seven minutes. Dr. Gibbon performed the operation.

The first successful hole-in-the-heart operation was performed on September 2, 1952, by Dr. F. John Lewis and Dr. Walton Lillehei (1918–99). The patient was a five-year-old girl.

The first heart bypass surgery was performed in 1967 by Argentinian cardiologist Dr. René Favaloro (1923–2000) in Cleveland, Ohio.

The first human heart transplant was performed by Christiaan Barnard (1922–2001), a cardiac surgeon at Groote Schuur Hospital in Cape Town, South Africa, on fifty-five-year-old retired grocer, Louis Washkansky (1913–67), on December 3, 1967. Washkansky died eighteen days later from pneumonia, not from failure of the new heart. The donor was Louise Darvall (1943–67), who had died after a car accident.

On January 23, 1964, three years before the first human heart transplant, Boyd Rush became the first human to receive a transplanted heart. While he was waiting for a human donor heart, Rush's own heart failed and in a last-ditch attempt to preserve Rush's life, the pioneering American surgeon James D. Hardy (1903–87) transplanted a chimpanzee's heart in its place.

The new heart immediately began to take over the role of the diseased heart, but within a few hours Rush's body rejected it.

ROBOT-ASSISTED HEART SURGERY

One of the most exciting areas of medicine currently being developed is the introduction of robot assistance. This technique developed out of the need to control the greater hand tremors experienced by surgeons when conducting operations using minimally invasive surgery.

In traditional heart surgery, the chest is opened and the surgeon is able to put his or her hands physically inside the cavity to make incisions close to the organ. In minimally invasive surgery, the incision may be no bigger than a few millimeters. The instruments are therefore much longer, and this added length magnifies normal hand tremors.

With robot assistance, the surgeon is also able to conduct the operation from a remote location, even from another country.

The first robot-assisted heart bypass operations were performed in late 1998 on seventeen patients by Dr. Ralph Damiano, at Pennsylvania State Hospital in the United States.

KIDNEYS

The artificial kidney (dialysis machine) was invented by Willem Kolff (b. 1911) during the Second World War, with the first machine being tested in 1943. In the great humanitarian tradition, Kolff refused to patent his invention. Kolff was born in the Netherlands but moved to the United States in 1950 and subsequently worked on development of an artificial human heart at his new home at the Cleveland Clinic Foundation.

Early experiments with kidney transplants began in France in 1909, with diseased human kidneys being replaced by animal kidneys. There were no survivors.

Before tissue and blood-type matches, and the human immune system, were fully understood, all human-to-human kidney transplants failed. Researchers began to realize that the body rejects anything it does not regard as its own, and a way was sought to combat those rejections.

In 1947, Charles Hufnagel (1916–89), a young surgeon working in Boston, in a desperate last-ditch attempt to save a patient's life transplanted a dead patient's kidney into the forearm of a young woman whose own kidney was so diseased that she had been given only hours to live. The woman was too weak to move to an operating theater, so the surgical team worked in her room with only rudimentary lighting to illuminate the operation. The kidney began to function as soon as it was connected to her blood supply. Although the transplanted organ died after a few days, it had provided sufficient breathing space for the woman's own kidneys to revive, and she made a full recovery.

The first successful kidney transplant was performed by Dr. Joseph Murray (b. 1919) on December 23, 1954, at the Peter Bent Brigham Hospital in Boston. He took a kidney from Ronald Herrick and transplanted it into his identical twin brother, Richard, allowing Richard to live for another eight years. Ronald remains alive today, as does Dr. Murray, who was awarded the Nobel Prize in Physiology or Medicine in 1990.

MIND

The first mental hospital in England was built in London in 1247 near to the present-day site of Liverpool Street Station. The hospital was named the Bethlehem Hospital, but was also known as Bethlem or Bedlam.

Mesmerism Franz Anton Mesmer (1734–1815), who was born in Austria, practiced his branch of medicine in Paris. Mesmerism, a form of hypnotism, was used in an attempt to treat mental illness.

The first mental hospital in the United States was the Institute of Pennsylvania Hospital, built in 1859.

Psychoanalysis was pioneered by Sigmund Freud (1856–1939). His first book, *Studies in Hysteria*, was published in 1895, and his seminal work, *The Interpretation of Dreams*, in 1899. Freud based his analytical practice on his theory of unconscious motives and the analysis of dreams, which were subjected to his "talking cure."

Before psychoanalysis, the principal treatment for mental illness was to restrain the patient physically, and often to subject him or her to a terrifying array of quack remedies such as surprise ice-cold showers; being kept in total darkness for days at a time to induce docility; and spinning around, strapped into a revolving chair, for long periods. Most treatments ensured that the patient (or victim) was never cured.

The first lobotomy for the relief of schizophrenia was performed in 1930 by Portuguese neurologist Egas Monitz (1874–1955). The procedure involved the insertion of medical instruments through the eye sockets to sever the nerves in the frontal lobe of the brain. He won the Nobel Prize in Physiology or Medicine for his development of the lobotomy operation and became the only Nobel laureate in the field of psychiatry.

Since the 1960s psychiatric drugs have made the lobotomy obsolete.

Electroshock treatment was first used in Italy in 1930 for cases of severe depression. With electroshock, typically one amp is administered, using up to 500 volts. Two thousand years earlier, Scribonius Largus (AD 14–54) experimented with neuro-stimulation by holding patients in water and allowing torpedo fish (electric rays) to administer severe electric shocks. Patients with gout experienced temporary pain relief.

Tranquilizers were first used in psychiatry in 1952 to restrain violent patients.

Prozac, the world's first and still most widely prescribed antidepressant drug, was launched by the American drug company Eli Lilly in 1988. The level of serotonin, a neurotransmitter in the brain that carries messages between nerve cells, is thought to influence appetite, aggression, and mood. Prozac works by increasing the levels of this chemical transmitter, and the drug was the first example of biochemistry being used to control mood.

Sex change (gender reassignment) In the modern era, the first American to undergo male-to-female sex change surgery was American GI George (later Christine) Jorgensen (1926–89). The surgery, which took place in Denmark, began in 1952 with the removal of the male organ, and after a long healing period, the process was completed in 1954.

TEETH

There was never yet philosopher
That could endure the toothache patiently.
—WILLIAM SHAKESPEARE (1564–1616)

In 1700 BC the Babylonians used gold and silver to make artificial teeth, and Egyptian mummies have also been discovered with false teeth. The pre-Roman Etruscans were able to affix bridges between teeth to fill gaps.

The first to advocate the use of fillings rather than extraction for partially decayed teeth was Scribonius Largus (AD 14–54) in AD 47. He made the dubious claim that the process of removing decay with the use of a knife was painless.

The first to use gold for fillings in teeth was Giovanni Arcolani, in 1493.

The first ivory false teeth were made around 1700 by German physicians, using mostly elephant and walrus tusks. However, even human teeth from cadavers were used.

Porcelain false teeth were introduced in France in 1774.

The first dentures set in a plate were produced by London silversmith Claudius Ash in 1845. The artificial teeth were set in an eighteen-karat gold plate, which incorporated springs and swivels to fit different mouth shapes. Ash established his business in 1829, and it continues to this day as part of the Plandent Group.

The first to recognize that nitrous oxide gas has anesthetic properties was the Cornishman Humphry Davy (1788–1825), in 1799. Since he was a research scientist, not a doctor, the medical profession ignored his findings until 1844.

The anesthetic property of cocaine injected into the gums was discovered in 1862 by Czech researcher Dr. Damian Schroff (1802–87). Because cocaine has serious addictive properties, research on synthesizing a substitute took place from 1880. An artificial substitute, novocaine (aka procaine), was formulated in 1905 by German chemist Alfred Einhorn. It has been used as a dental anesthetic ever since.

DRUGS

PENICILLIN

The first person to study the antimicrobial capabilities of penicillin mold was the French physician Ernest Duchesne (1874–1912), in 1897, while working at the school of the military health service in Lyon. Despite his meticulous experiments, the Institut Pasteur, noting Duchesne was only twenty-three, ignored his findings. In 1904 Duchesne contracted tuberculosis, and he died in 1912 without his findings progressing any further.

The antibiotic and antiseptic properties of penicillin were first discovered by Alexander (later Sir Alexander) Fleming in 1928. When he

presented his results to the Medical Research Club, Fleming was horrified, but not initially discouraged, when his colleagues ignored his low-
key presentation of "a powerful anti-bacterial substance."

The establishment continued to ignore Fleming's findings, and it was
in 1939, when he attended the Third International Congress of Microbi-
ology in New York, that he found American researchers were also working on penicillin. It was only in 1940 that the medical profession, under
pressure from the effects of the Second World War, fully accepted penicillin as an antibiotic.

Industrial manufacture of penicillin began in 1942.

ASPIRIN

As early as 1829 scientists Johann Buchner of Munich and Henri Ler-
oux of France had discovered that an extract from the willow tree called
salicin could provide pain relief for headaches. However, the problem with
salicin was that it caused stomach inflammation, and could cause vomit-
ing of blood. In 1853 a German chemist, Charles Gerhardt (1816–56),
found a way of preventing stomach inflammation by mixing the com-
pound with sodium, but he had no desire to market his discovery.

In 1897 Felix Hoffman (1868–1946), working for the Bayer chemical
company in Germany, rediscovered Gerhardt's formula and produced
the product in powder form. Bayer created the trademark Aspirin and
marketed it as a powder until 1915, when aspirin in the form of tablets
were first put on sale. The trademark Aspirin was lost by Bayer as part
the reparations Germany was forced to pay under the terms of the 1919
Treaty of Versailles following its defeat in the First World War.

VIAGRA

The American drug company Pfizer developed Viagra, initially as a
treatment for angina. In clinical trials it failed to display any benefits for
angina, but reports noted that it had the marked side effect of inducing
strong sexual arousal in the male subjects. Viagra first became available
in the United States in 1998 and in the United Kingdom in 1999. Originally, it was only available on a doctor's prescription, and within three
years of being launched, sales of Viagra had topped $1 billion worldwide.

INVENTIONS

*Today, every invention is received with a cry
of triumph, which soon turns into a cry of fear.*
—BERTHOLT BRECHT (1898-1956)

However useful an invention may seem right now, and however difficult it may appear to be able to improve on it, someone somewhere will either find a way to improve it, or produce a different solution. It is in the nature of men and women to seek ways to improve the way life is lived, and throughout history the products of their combined imaginations and experiments have brought tidal waves of inventions.

Some inventions such as televisions, vacuum cleaners, motorcars, or corkscrews become everyday objects. Some, like the Claxton earcap, don't.

Some inventions, like computers, telephones, airplanes, and the artificial heart have significantly improved life for the population of the world. Others, like the tobacco resuscitator, have more in common with implements of torture. They should be used on the inventor first to see if he likes it. Here are just a few inventions and their origins.

COMPUTERS

There is no reason why anyone
would want a computer in their home.
—KEN OLSEN (B. 1926)

There are numerous claims as to what was the first computer. Here are some:

THE ABACUS

A device used for mechanical addition, subtraction, multiplication, and division, the abacus does not require the use of pencil and paper and is good for any base number, but normally in groupings of ten.

There are two forms of abacus: one uses counters on a board with special markings, the other uses beads strung on wires fixed into a frame. The counters or the beads are used to assist with addition, subtraction, multiplication, and division, by using one set of beads to count each of the tens, hundreds, and thousands.

It is thought that the abacus was invented by the Babylonians and may have been used as early as 2400 BC, but the more commonly accepted dates are between 1000 and 500 BC. There is also evidence of a similar Chinese invention as early as 3000 BC, and the Aztecs had a form of abacus from about 1000 BC. The abacus was in common use by the Japanese up to the 1920s.

Mechanical computers

The Antikythera mechanism was built in 87 BC by an unknown craftsman to calculate the new moons. It had a specially designed gear ratio of 235 to 19.

A mechanical calculating machine was designed by Leonardo da Vinci (1452–1519) in 1500.

A logarithmic calculating device was developed in 1620 by Edmund Gunter (1581–1626) of England. This is widely regarded as the first successful analog device. Analog computers perform operations in parallel steps, meaning they can perform more than one operation at a time.

The slide rule was invented in 1621 by the clergyman and mathematician William Oughtred. It is another version of an analog computer and consists of three interlocking strips, normally wood or plastic, which are calibrated so that positioning the strips relative to each other enables arithmetical calculations to be made.

The first mechanical digital calculating device was built by Blaise Pascal (1623–62) of France in 1642.

A "difference engine" was built by the English mathematician Charles Babbage (1791–1871) in 1822 to improve significantly the accuracy of the calculations in the production of arithmetical tables. Difference engines had first been conceived in 1786 by J. H. Mueller of Germany but did not leave the drawing board.

Babbage's fantastic and complex machine is considered to be the most beautiful computer ever built. It works on the principle that it only ever needs to be able to subtract numbers. The British Museum possesses a working model of Babbage's difference engine.

Electronic computers

Digital computers work on the principle of binary code, using only the figures zero and one for all calculations, storage, retrieval, and instruction.

The modern computer came into being in 1939 when the Bulgarian-American physicist John Vincent Atanasoff (1903–95) built the first electronic digital computer. Atanasoff, whose father had emigrated from Bulgaria to the United States in 1889, hired Clifford Berry, a young electrical engineer, to help him and they named their first machine the ABC (Atanasoff-Berry Computer).

The first electronic brain was built by Alan Turing (1912–54) at Bletch-

ley Park during the Second World War, in order to crack the almost unbreakable code of the German Enigma cipher machines. It became known
as the Turing-Welchman Bombe, and it made the rapid calculations needed
to solve the immensely complex Enigma codes successfully. Turing's great
work helped to shorten the Second World War in Europe. (See War, p. 272.)

The algorithm was invented by Alan Turing before the Second World
War began. In late 1936, Turing had published a paper, "On computable
numbers with an application to the Entscheidungs problem." With this
paper Turing effectively invented the algorithm, which is a set of instructions for accomplishing a task. This in turn led to the conception of Turing's machine, and has now become the foundation of modern
computing and the stored computer program.

A fully automatic large-scale calculator was built in 1944 by
Howard Aiken (1900–73) of the United States. It was known as the Harvard Mark I, had more than 750,000 parts, and was reputed to sound like
a roomful of ladies knitting.

The first programmed electronic computer was built in 1946 by
J. Presper Eckert (1919–95) together with John W. Mauchly (1907–80). It
was named the ENIAC (electronic numerical integrator and calculator)
and contained 20,000 vacuum tubes.

The first stored program computer was the EDVAC (electronic dis-
crete variable automatic computer). It was developed during the late
1940s and introduced in 1952. This followed the definitive paper on the
subject, entitled "The First Draft," written by Hungarian-born mathematician John von Neumann (1903–57).

The semiconductor or integrated circuit was invented by Robert
Noyce (1927–90) of Fairchild Semiconductor and Jack Kilby (1923–2005)
of Texas Instruments, who were working separately and without knowledge of the other's work. Kilby patented the discovery in 1958 and won
the Nobel Prize for Physics in 2000. The invention stands as one of the
most important of the twentieth century. Transistors, resistors, capaci-
tors, and all the associated connecting wires were incorporated into a single miniaturized electronic circuit. The two companies shared information
and helped to create a trillion-dollar industry.

Noyce went on to found Intel, the company that developed the computer microprocessor.

*What we didn't realize then was that the integrated circuit
would reduce the cost of electronic functions by a factor of a
million to one. Nothing had ever been done like that before.*
—JACK KILBY (B. 1923)

The computer microprocessor was invented in 1968 by U.S. engineer Marcian "Ted" Hoff (b. 1937). It is also called a microchip or chip, which places all the thinking parts of a computer, such as the central processing unit (CPU) and the memory, onto a single silicon chip.

Hoff joined Intel as employee number twelve, and his invention was first marketed in 1971 as the Intel 4004. Eighty percent of the world's computers now operate on Intel microchips.

In 1949 Edmund Berkeley (1909–88) published plans to build SIMON, in his book *Giant Brains, or Machines That Think*. SIMON was a desktop machine, about 4 cubic feet in size, that used relay technology. Some experts regard it as the forerunner of the personal computer.

In the 1970s and early 1980s the terms *home computer, microcomputer, desktop computer,* and *personal computer* were interchangeable. There are debates on the first true personal computer, but the Xerox Alto, with a graphic interface, which was launched in 1974, has a strong claim. So too does the Altair 8800, launched in 1975, which was the first computer to carry the Microsoft programing language, BASIC.

The first personal computer sold with an integrated keyboard and monitor, was the Apple 1, which was introduced by Apple in 1976. The casing was made of wood.

PARACHUTES

The principle of how parachutes would work was recognized by Leonardo da Vinci as early as 1480, but the first practical demonstration was conducted in 1783 by French physicist Louis-Sébastien Lenormand (1757–1839) when he jumped from a tree holding two parasols, landing successfully. There are accounts of a similar experiment in China in 90 BC.

The first parachute jump proper took place in 1797 when Parisian

André-Jacques Garnerin (1769–1823) jumped successfully from a balloon, it is said from approximately 3,200 feet, using a parachute with a basket slung underneath. Garnerin's wife, Jeanne-Geneviève, became the first woman parachutist in 1799. In 1802, Garnerin made the first parachute jump in Britain, from 8,000 feet.

The first successful parachute descent from an airplane was in 1912 by U.S. Army Captain Albert Berry (1852–99).

Parachutes were not issued to pilots in the First World War. Aircraft were in shorter supply than men, and High Command held the theory that pilots might take the easy way out of a dogfight and bail out, rather than try to save the plane. Hundreds died in burning planes, or bailed out without parachutes and fell to certain death.

The "sail" type of steerable parachute was developed during the Vietnam War in the late 1960s so that pilots bailing out at high altitude had a chance to steer themselves back behind their own lines.

NEWCOMEN ENGINES

These atmospheric steam engines, which were designed to pump water from deep mines, were invented by Thomas Newcomen (1663–1729). The first engine was installed in 1712 close to Dudley Castle in the English Black Country.

Newcomen had to share his success with Thomas Savery (1650–1716), who had previously taken out a general patent covering all the possible means of pumping water by steam power.

PHOTOGRAPHY

Predating photography, and known in China as early as the fifth century BC, the camera obscura (from the Latin meaning "dark chamber or room") was originally used to observe eclipses of the sun without damaging the eyes. The mechanism consists of a darkened room into which light enters through a small hole, casting an inverted image on to the opposite wall or onto a screen.

By the sixteenth century the camera obscura was being used as an aid to drawing. Eminent artists such as the Italian Giovanni Antonio Canale (1697–1768), better known as Canaletto, used the camera obscura to observe images very closely and then produce highly accurate paintings of buildings.

Photography was created when French inventor Joseph Nicéphore Niépce (1765–1833) introduced light-sensitive paper into a camera obscura. In 1816 Niépce managed to record a view from his workroom on paper that had been sensitized with silver chloride, but was only able to partially fix the image. It took him until 1826 to make a permanently fixed image on a pewter plate. The exposure took eight hours, and the resulting photo is now held in the library of the University of Texas. Niépce called his discovery heliography (from the Greek for sun drawing) since his images resulted from the exposure of the paper to sunlight.

Niépce also invented a means of extracting sugar from beetroot, as well as a machine that he called a pyréolophore, which used the expanded air in a controlled explosion to create propulsion. The pyréolophore was a forerunner of the internal combustion engine (see "Questionable Origins," p. 187).

A primitive form of photography was already being used in 1839 by Sir John Herschel (1792–1871), the same year that Louis Daguerre announced the first commercial photographic system: the daguerreotype.

In 1841 William Henry Fox Talbot (1800–77), based in London, patented the more commercial negative–positive system, which is the basis of the modern photographic process.

Flexible film was invented in 1884 and introduced in 1889 by George Eastman (1854–1932) of Eastman Kodak fame.

Color photography was successfully demonstrated at the Royal Institute in London in 1861 by James Clerk Maxwell. Early color photography was extremely complex and prohibitively expensive. In 1869 Charles Cros and Louis Ducos du Hauron devised a method of creating color photographs.

The electrically ignited flashbulb was invented in 1893 by E. J. Marcy.

The first commercially available digital camera was the DCS 100. It was released in 1990 by Kodak and was aimed at the professional market. In January 2004, Kodak announced that production of nondigital cameras would cease at the end of the year.

SEWING MACHINES

The first mechanical device for sewing was patented in 1755. British Patent 701 was granted to Charles Frederick Weisenthal of Germany.

The first sewing machines in the United States were produced in 1839 by Lerow and Blodgett Inc. After repairing a Lerow and Blodgett machine, Isaac Merit Singer (1811–65) could see ways to improve the design and patented his own machine in 1851. He formed the Singer Sewing Machine Corporation, which became a worldwide company.

Singer also pioneered payment plans so that a sewing machine could be taken away for a five-dollar downpayment. He retired to England in 1863.

The earliest sewing machine for mass-production of clothing was designed by Barthélémy Thimmonier of France in 1841. It was designed for the production of French army uniforms but was destroyed by rioting tailors.

BUNSEN BURNERS

Most school laboratories are equipped with Bunsen burners, which are used to heat liquids in test tubes. The combustible fuel is a mixture of gas and air, and the temperature is regulated by a simple slide to control the flow of air into the combustion tube.

Bunsen burners were first produced in 1855 by German chemist and physicist Robert Bunsen (1811–99). The original designs were by either Michael Faraday or Peter Desdega.

CASH REGISTERS

The cash register was invented in 1879 by bartender James Ritty (1837–1918). He opened a factory, which he called the National Manufacturing Company. In 1884 Ritty sold the business to James Patterson for $6,500.

The name was changed to National Cash Register, and the company became a major international business, a forerunner of the early computer industry, and a subsidiary of the massive American Telephone and Telegraph Corporation in 1991.

RADAR

In 1887 German physicist Heinrich Hertz (1857–94) demonstrated the existence of radio waves. He also demonstrated that they behave like light waves and can be reflected. (See "War," p. 261.)

A device using radio echo for marine navigation was patented in 1904 by German engineer Christian Hulsmeyer.

Practical methods of using radar for aircraft detection were developed in 1935 by Sir Robert Watson-Watt (1892–1973), a descendant of James Watt. By 1939 he had installed a chain of radar stations across the south and east coasts of England.

During the Second World War, German High Command was concerned about high losses to their night fighters and bombers, thanks to the British radar detection systems. British counterespionage put it about that Allied pilots were eating large quantities of carrots every day to improve their night vision. The secret of RAF onboard radar was protected for several months.

SONAR AND ASDIC

Sonar (sound navigation and ranging) and asdic (anti-submarine detection investigation committee) are identical systems of underwater detection of solid objects by means of sonic echoes. Both active listening devices were developed during the Second World War: sonar in the United States, and asdic in Britain.

ELECTRICITY AND
ELECTRICAL APPLIANCES

Thomas Alva Edison (1847–1931) was the pioneer who brought the world's first reliable, low-priced, public electricity supply to New York in 1882. This led to the development of electrical appliances that could be mass-produced for the domestic market.

DISHWASHER

American Joel Houghton developed the first dishwasher in 1850. He patented a wooden machine with a hand-turned wheel, which splashed water on dishes. It failed to become a household item.

In 1886 Josephine Cochrane, a wealthy widow in Illinois, invented and patented a dishwashing machine because she was dissatisfied with the treatment her servants were giving to her china and disliked doing the washing-up herself.

After her husband died, she launched a commercial venture with her machine, which was capable of washing a load of dishes in two minutes. Basically, it consisted of wire compartments, into which the dirty china was placed. These were then put onto a wheel inside a copper boiler. The wheel was turned by a motor as hot, soapy water cleaned the china. The machine was used in hotels and restaurants but not in family homes.

KETTLE

The electric kettle was launched in England by Crompton and Company in 1891.

TOASTER

General Electric introduced what it claimed was the first electric toaster in 1909. A competing claim by Hotpoint puts its product launch date at 1905.

The pop-up toaster was invented in 1919 (patent granted in 1921) by American Charles Strite and launched in 1926.

FOOD MIXER

RAF engineer Kenneth Wood (1916–97) had the idea of a multipurpose food mixer and launched his first product, the A200, in 1947. He launched the famous Kenwood Chef in 1950.

AIR-CONDITIONING

A form of air-conditioning has been known in India for hundreds of years. Wet leaves are draped across the entrance to a building. As air currents pass into the building, the air is cooled by the water evaporating from the leaves. The principle of exchanging cool air for hot air is the same in the modern air conditioner.

The modern electric air conditioner was invented by New Yorker Willis Haviland Carrier (1876–1950) in 1902. At the time, he was employed at a printing company, which was experiencing problems with four-color printing due to inconsistencies in the air quality. Carrier came up with a solution and was awarded a patent in 1906.

Air-conditioning for human comfort, as opposed to being part of an industrial process, began in 1924 at the J. L. Hutton Department Store in Detroit, Michigan.

VACUUM CLEANER

A forerunner of the vacuum cleaner was invented in 1901 by British engineer H. Cecil Booth. He developed a gasoline-powered, horse-drawn cleaning device, which he called Puffing Billy. The device was for commercial, not domestic, purposes, and would be parked outside office buildings and shops, with a long hose running inside for the cleaning operation. It was not a commercial success.

The domestic vacuum cleaner as we know it was invented in 1907 by James Spangler, a janitor in Ohio. Dust was thrown up by the carpet sweeper he used in his job and continuously troubled his asthma, so he needed to trap the dust. He connected a fan motor to a broom handle, collected the dust in an old pillowcase, and was awarded a patent.

William Hoover (1849–1932), the husband of a customer of Spangler, liked the product so much he bought the company and rebranded the product line Hoover. The name Hoover became the generic term for a vacuum cleaner, and hoovering the verb to describe the action of vacuuming. The company went on to become a worldwide industrial combine.

REFRIGERATOR

In ancient India, Egypt, Greece, and Rome, wealthy citizens made use of snow cellars, which were pits dug into the ground and filled with straw and wood. Ice transported from mountains could be stored for several months in this way.

The first domestic refrigerator was developed in 1834 by American inventor Jacob Perkins (1766–1849). The machine worked by manually activating a pump and converting the heat produced into a heat loss between two separate chambers. It was not a commercial success because of the exceptionally long time it took to reduce temperature sufficiently to cool liquids and keep food fresh.

The first commercially successful domestic refrigerator was produced in late 1916 by the Kelvinator Company of Detroit, which is now owned by Electrolux.

LIGHTBULB

In 1811 the Cornish scientist Sir Humphry Davy (1778–1829) discovered that light is produced when passing an electric arc between two poles. By 1841 Davy's electric arc lights had been installed in Paris.

In 1879 Sir Joseph Swan (1828–1914) of the United Kingdom and Thomas Alva Edison (1847–1931) of the United States simultaneously, on opposite sides of the Atlantic, invented the electric incandescent lightbulb.

PHOTOCOPYING

Known more accurately as xerography, photocopying was invented in 1937 by Seattle-born Chester Carlson (1906–68). It works by charging paper and powdered ink with opposite static electrical charges. By applying heat to the whole product the ink is fixed to the paper. The massive commercial potential was not realized until the 1950s when photocopying was exploited by the Haloid Corporation, later renamed Xerox.

The first color copiers were developed in the 1970s.

SMOKING

If I cannot smoke in heaven, then I shall not go.
—MARK TWAIN (1835–1910)

Experts believe that tobacco was originally used as a hallucinogenic enema in the Americas around the first century BC. The Maya, who occupied the Yucatán Peninsula in present-day Mexico, developed tobacco as a smoking material between AD 470 and 600, and tobacco was also used by the Mississippi Indians around the same time.

The first pictorial evidence of smoking is on an eleventh-century pottery vessel found in Guatemala, showing a Maya smoking a roll of tobacco leaves tied with string.

The first Europeans to observe smoking, in November 1492, were two Spaniards, Rodrigo de Jerez and Luis de Torres, who were present on Christopher Columbus's (1451–1506) exploration that first sailed to the Americas.

Jerez is thought to be the first non-American smoker. When he returned to Spain, he was imprisoned for seven years by the Inquisition having been accused of frightening people with the smoke billowing from his mouth and nose. By the time of his release, smoking was widespread in Spain.

Tobacco was introduced to England in 1564–65 by Sir John Hawkins (1532–95). It was known as "sotweed" in Elizabethan times.

In 1585 Sir Francis Drake (1540–96), a cousin and protégé of Sir John Hawkins, showed Sir Walter Raleigh (1552–1618) how to smoke tobacco using a clay pipe.

Tabacco, the first English book on the subject of tobacco, was published in 1595.

In 1604 (published in 1672) King James I of England wrote "A Counterblaste to tobacco," which strongly denounced the use of tobacco, calling it "lothsome to the eye, hatefull to the Nose, harmfull to the Braine, dangerous to the Lungs . . ." and in 1629 the first No Smoking signs were being put up in some government buildings in Britain.

In 1609 John Rolfe began to raise tobacco on his Jamestown land and became the first person to successfully export tobacco from Virginia to Europe. The strain of tobacco used by Rolfe (*Nicotiana tabacum*), was imported from Bermuda with the specific intention of growing a crop that would appeal to the European palate. The export income from tobacco as a cash crop helped to establish the permanency of the American colony.

Consistent production of yellow-colored tobacco, which was milder than the more usual dark brown variety, was begun in 1856 by Abisha Slade of Caswell County, North Carolina, who developed an industrial means of adding charcoal to the curing process. The process had been accidently discovered in 1839 by Stephen, a slave who belonged to the Slade family.

The first clinical study into the effects of using tobacco in all its forms was made as early as 1761 by the London physician Sir John Hill (1716–75). Hill linked cancer with tobacco use and noted the high incidence of cancers of the nose in snuff (powdered tobacco) users.

The cigarette was invented in 1832 by an Egyptian artilleryman at the Battle of Acre in the Turkish-Egyptian war. The gunner had managed to improve the rate of fire of his cannon by rolling the gunpowder in paper tubes. He was rewarded with a supply of tobacco. Having broken his pipe, he rolled the tobacco in the same paper that he had used for the gunpowder and smoked what became the first cigarette.

In 1877, in an attempt to mechanize the production of cigarettes, the Allen and Ginter Tobacco Company of Richmond, Virginia, offered a prize of $75,000 to anyone producing a successful cigarette-making machine. In 1880, eighteen-year-old James Bonsack developed a machine, but after installation in the Allen and Ginter factory, it failed after several trials. In 1884 W. Duke Sons and Company installed a modified Bonsack machine, which, when it worked, could do the work of forty-eight hand rollers.

At the instigation of Buck Duke, the five principal tobacco manufacturers merged in 1890 to avoid competing against each other. They formed the American Tobacco Company, which controlled 90 percent of the U.S. tobacco market, and was by far the largest tobacco company in the world. The company became known as the Tobacco Trust and was seen to be working against the public interest. Antitrust legislation was introduced, and the company was broken up in 1911. Buck Duke went on to found Duke University.

The first cigar-rolling machine was patented in 1883 by the composer Oscar Hammerstein.

SAFETY PINS

The humble safety pin was the creation of Walter Hunt (1785–1869) of New York. Hunt's wife complained continuously about pricking her finger with pins that were available at the time, which were straight and had to be bent by hand. In 1849 Hunt decided to do something about it. He fashioned a piece of brass wire into the shape we are familiar with today, with a simple catch to shield the sharp end, and was granted a patent.

Unfortunately for Hunt he was badly in debt at the time and was forced to sell his patent for just four hundred dollars.

CORKSCREWS

The apparently simple requirement of removing corks from bottles has inspired armies of inventors, who have come up with an almost limitless array of eccentric devices. These range from swords, which were used to chop the necks from champagne bottles, to handheld compressed-air hypodermic needles, for forcing corks out under pressure.

The ancient Romans used cork to seal containers of wine, but the design of the wine bottle with the narrow neck is attributed to Englishman Sir Kenelm Digby (1603–65). Digby, whose father was executed for involvement in the Gunpowder Plot, invented the narrow neck bottle in about 1630, and provided waxed linen wrappings around the corks, enabling them to be extracted with no mechanical assistance. To allow horizontal storage, the design of wine bottles evolved, during the seventeenth century, with longer and narrower necks. With bottles being stored horizontally, corks were now in permanent contact with the liquid, and had to fit ever more tightly.

To achieve the best fit, corks were compressed and became, in the process, more difficult to remove. The first corkscrews were produced in the late seventeenth century and were based on the earlier "gun worm," which is a helical device for removing unspent charges from a musket's barrel.

With his eye firmly fixed on the ready market for a reliable product, enter, in 1795, the Reverend Samuel Henshall from Oxford, England. His design became the world's first patented corkscrew. The idea behind the Henshall patent was to have a concave disk on the shank of the screw, which limited the distance it could penetrate into the cork. By continuing to turn the handle once the limit had been reached, the cork was compressed further, thereby breaking the bond between cork and glass.

Perhaps the most unusual bottle-opening method was used by eighteenth-century French cavalrymen who bolstered their courage by riding into battle armed with a saber and a bottle of champagne. Before beginning a cavalry charge, they would "decapitate" the bottles using their sabers. The resulting explosion of champagne removed any shards of glass, and they would drink directly from the open bottle.

USELESS INVENTIONS

Laughable they may be, but useless inventions serve to remind us of what might have been.

CODPIECE

This was worn in fifteenth-century Europe by dandyish males to support their genitalia. Sometimes the codpiece would be decorated with jewelry and generally enhanced to look prominent. By 1580 they had passed out of fashion since they were thought to look lewd.

CHASTITY BELT

The chastity belt was designed in the Middle Ages to prevent married women from engaging in sexual intercourse outside of marriage. The belts were normally fitted to the women during the absence of their husbands on the Crusades. They were unpopular and uncomfortable.

CLAXTON EARCAP

Adelaide Claxton patented the Claxton earcap in 1925, which was supposed to be worn at night by children to correct ears that stuck out.

PHRENOLOGY

This is the so-called science of determining a person's character by the shape of his or her skull. Franz Joseph Gall (1758–1828) promoted phrenology in the late eighteenth and early nineteenth centuries.

TOBACCO RESUSCITATOR KIT

Invented in England in 1774, the tobacco rescuscitator kit was meant to revive victims of drowning, by injecting tobacco smoke into the rectum.

The idea was to introduce warmth and stimulation to the apparently dead victim, and this was thought to be an entirely rational development since tobacco was effective in providing both. The devices were placed along the banks of London's river Thames by the Royal Humane Society, in the same way that flotation rings are placed today.

YARD-OF-ALE GLASS

The "yard-of-ale" glass is a peculiar British drinking vessel that was first used in the seventeenth century. The glass has a trumpet-shaped opening, a long neck of about a yard in length, and a bulbous container. As the glass is tipped up, and the drinker begins to empty the contents, there is a rush of air, which propels the contents toward the opening at a far-increased speed, creating a wet drinker, to the amusement of on-lookers.

The difficulty of drinking from it renders the yard-of-ale glass imprac-tical for normal use, and they are mainly used for betting.

LANGUAGE

COVERING EARLIEST ORIGINS, INDO-EUROPEAN,
GREEK, LATIN, ENGLISH, SIGN LANGUAGE, BRAILLE,
MATHEMATICAL SYMBOLS, AND SLANG

*In Paris, they simply stared when I spoke to them in French.
I never did succeed in making those idiots understand
their own language.*
—MARK TWAIN (1835–1910)

EARLIEST ORIGINS

Until the eighteenth century and the Enlightenment (a European intellectual movement), most thinking about the origin of language
assumed it began with Adam and Eve in the Garden of Eden. The most
recent theory of the origin of language is that simple hand gestures were
used as long ago as 6 million or 7 million years, shortly after the human
line diverged from the apes. Shouting would have been used for alarm
calls or emotional outbursts.

About 5 million years ago, an early hominid known as *Australopithecus*
started to walk upright, and a more complex form of gesturing was probably used.

Then, 2 million years ago, brain size increased and hand gestures were used in various combinations to express ideas, and remained the primary means of communication.

As recently as 100,000 years ago, Homo sapiens may have changed the main means of communication from hand and facial gestures to vocalizations and the use of differing sounds to convey a variety of meanings. Gradually, gesturing diminished, although we still use it today to emphasize speech, even during telephone conversations when the person at the other end cannot see the gestures.

THE FIRST WORD

In the beginning was the word.
—JOHN 1:1

After studying a thousand languages, researchers in Paris have come to the conclusion that *papa* was the first word ever used. They discovered *papa* in seven hundred of the languages studied. From the start of the last Ice Age, roughly 50,000 years ago, it seems that the word *papa* was part of a common language.

OLDEST SPOKEN LANGUAGE

The oldest of all spoken languages is almost certainly Mayan, which seems to have been in existence seven thousand years ago when the Maya migrated south into Mexico.

There are thirty Mayan languages spoken today, which are so closely linked that linguists believe they all originated from a single proto-Mayan language.

FIRST WRITTEN LANGUAGE

Cuneiform, the first written language, was developed by the Sumerians more than five thousand years ago. The strange wedge-shaped letters were a development from earlier pictograms and were formed by pressing shapes into wet clay with a special stylus. Cuneiform was the first language able to convey abstract ideas and sounds. There were only two numbers in cuneiform, a vertical wedge for one and a horizontal wedge for ten.

INDO-EUROPEAN

The original Indo-European language, known as proto-Indo-European, was used around five thousand years ago. Because it did not develop as a written script, historians are not able to pinpoint the precise area in which proto-Indo-European arose as a spoken language, but it is thought most likely to have come either from the inhabitants in an area of present-day Poland or Turkey.

Around three thousand years ago Indo-European split into two main groupings, the Germanic and Romance groups. In the second century BC the Germanic group split further into East, North, and West subgroups, of which West Germanic is the ancestor of English.

GREEK

Although the Greek language is widely regarded as the mother tongue of Western civilization, its beginnings are modest in scale compared with Mesopotamian and Egyptian. Minoan civilization developed on the island of Crete in about 2000 BC, and with it developed the beginnings of the Greek language. Minoan civilization was brought to an end around 1400 BC following the devastating effects of the volcanic eruption on the island of Thera and subsequent conquest by Mycenaean invaders from the Greek mainland.

LINEAR A

The early form of writing, Linear A, is thought to have been a written script rather than a spoken language. It replaced hieroglyphics in Minoan writing between 1700 and 1600 BC and although it has only ninety symbols, Linear A has resisted all attempts to decipher it.

ΑΒΓΔ

LINEAR B

Although it is by no means certain, Linear B may have developed out of Linear A. It was in use from 1500 to 1150 BC and was used for writing Minoan Greek. It is thought to be up to five hundred years older than the language used by Homer. Linear B was deciphered in 1952 by Michael Ventris (1922–56).

CLASSICAL GREEK

From around 1200 BC the Dorians migrated down from northwest Greece to occupy central Greece. The remaining vestiges of Mycenaean and Minoan culture were swept away. The country fell into its own Dark Ages, during which writing disappeared, and the Greeks appear to have become illiterate.

Around 750 BC, classical Greece emerged as a collection of city states, with heavy involvement in maritime trade, literature, politics, and philosophy. It reached its zenith in the fifth century BC.

In 338 BC, Philip II of Macedon (382–36 BC) had established Macedonian supremacy in Greece. His son, Alexander the Great (356–23 BC), spread Greek culture and language to other countries.

MODERN GREEK

The modern Greek language has the longest continual existence of any of the Indo-European languages, having begun to emerge in the fourteenth century BC and continued through to the present day. The language evolved came from ancient Greek then Koine, which was the common language from the fourth century BC to the fourth century AD, then on through Byzantine Greek from the fifth to the fifteenth century. Greek as we know it today has been spoken since the fifteenth century. Linguists place the actual starting date at 1453, the fall of Constantinople to the Ottoman Turks.

THE GREEK ALPHABET

This was the first recognizably modern alphabet, the revolutionary feature being the use of five letters to signify the five vowel sounds. Both the

Roman alphabet, now used throughout western Europe, and the Cyrillic alphabet, which is used in the six Slavic languages (Russian, Bulgarian, Serbian, Ukrainian, Belarusian, and Macedonian), are based on the Greek model.

LATIN

The Romans would never have had time to conquer the world
if they had been obliged to learn Latin first of all.
—HEINRICH HEINE (1797-1856)

An Indo-European language, Latin is the ancestor of the modern Romance languages: Italian, French, and Spanish. It was first spoken from the seventh century BC by small groups of people living along the banks of the river Tiber. The use of Latin rapidly spread as the Roman Empire conquered Europe and the Mediterranean coasts of Africa.

Latin was the most widely used scholarly language in Western civilization until the late Middle Ages and was the language of the Roman Catholic liturgy until the late twentieth century. During the classical period there were three types of Latin in use: classical written Latin; classical oratorical Latin, used in public speaking and prayer; and colloquial Latin as used in everyday speech.

THE ROMAN (LATIN) ALPHABET

This alphabet is now the most widely used writing system in the world, including, as it does, English and most of the European languages. It was developed around 600 BC from the Etruscan alphabet and can be traced right back to the North Semitic alphabet that was in use in Syria and Palestine around 1100 BC. The earliest surviving example of the Roman alphabet in use is on a seventh-century BC cloak pin. It says, "MANIOS MED FHEFHAKED NUMASIOI," which means "Manius made me for Numerius." The pin was probably a little love token.

There were only twenty-three letters in the Classical Latin alphabet.

During medieval times the letter *I* was used for both *I* and *J*, and the letter *V* was used for *U*, *V*, and *W*. Hence the twenty-six letters in the modern English alphabet.

Roman and italic typefaces, or fonts, commonly used in modern printing, developed from legal and commercial script used in fifteenth-century Italy.

ENGLISH

English is a member of the Indo-European family of languages, which includes most modern European languages (see above).

OLD ENGLISH

Also called Anglo-Saxon, there are four known distinct dialects: Northumbrian (northern England and southeast Scotland), Mercian (central England), Kentish (southeast England), and West Saxon (southern and southwest England). It is thought to have been spoken by King Alfred in the ninth century. The language had three genders—male, female, and neuter—and was spoken from around AD 500 to 1100. Anglo-Saxon dominated the Celtic speakers who were pushed westward into Wales, Cornwall, and Ireland and northward into Scotland. The Viking invasions, beginning in AD 850, added many new words and roughly 15 percent of modern English can be directly traced back to Old English.

MIDDLE ENGLISH

The Norman Conquest, beginning in 1066, introduced Old French (also known as Anglo-Norman). This developed into Middle English, which was in use from around 1100 to 1500. Middle English of the late fourteenth century would be recognizable today as English.

EARLY MODERN ENGLISH

William Shakespeare (1564–1616) strongly influenced early Modern English (1500–1800). He not only recorded certain words for the first time, but also invented words such as *dwindle* and *leapfrog*. In total, there are around two thousand words that can be attributed to Shakespeare.

LATE MODERN ENGLISH

The many new territories held in the British Empire, which at its peak covered a quarter of the globe, influenced late Modern English (1800–present). For instance, words such as *bungalow*, *shampoo*, and *jodhpur* are Indian. New technologies have brought new words to describe things that hadn't previously existed; words such as *vaccine*, *nuclear*, *typewriter*, and *microchip*.

AMERICAN ENGLISH

> *England and America are two countries*
> *divided by a common language.*
> —GEORGE BERNARD SHAW (1856–1950)

The English spoken in America began to change immediately after colonization began in around 1600. The American lexicographer Noah Webster (1758–1843) compiled dictionaries and what he called "spelling books" specifically to help Americans adopt their own standard of language, and to create uniformity. Webster's books were published between 1783 and 1828, and he named the language that he was institutionalizing "Federal" English.

There are three periods in the development of American English: Colonial (1607–1776), National (1776–1898), and International (1898–present).

SIGN LANGUAGE

It is almost certain that simple sign language, in the form of shoulder shrugs, pointing with the arms and fingers, and facial grimaces, is older than speech itself.

The first form of sign language known to have been specifically created for the deaf was developed by the Italian Giovanni Bonifacio in 1616.

After a chance encounter in the 1750s with two deaf sisters, who managed to communicate with each other by hand signals, Charles Michel, Abbé de l'Épée (1712–89) developed sign language in its modern form, as a form of communication for the deaf for spelling French words. French Sign Language (FSL) was also able to express whole concepts with a single sign, and remains in use today. De l'Épée came from a wealthy Versailles family and funded the world's first free school for deaf children from his own money.

American Sign Language (ASL) was developed in 1816 from FSL when it was taken to the United States. ASL is now the fourth most common language in America.

Deaf sign languages are independent of their spoken counterparts. The British and American sign languages are so different that one is almost unintelligible to users of the other language, and yet their spoken counterparts are almost identical.

BRAILLE

A form of written communication, Braille uses raised dots on paper or metal to enable blind people to read.

In 1821 French Army captain Charles Barbier invented what he called "night writing." The invention made use of a twelve-dot code that had been developed for battlefield conditions specifically to eliminate the need for speaking during the hours of darkness, when silence was vital. To demonstrate his new code, Barbier visited a school for blind boys, at which one of the pupils was Louis Braille (1809–52).

Braille was very interested in the system, but considered it too complex, so he set about simplifying it down to six dots. The dots were arranged in different positions to represent the letters of the alphabet, and the Braille system has now been adapted to almost every language in the world.

Braille, who was born near Paris in 1809, had accidentally blinded himself in one eye at the age of four while playing with an awl (a sharp-pointed instrument for marking surfaces or punching small holes). He

lost the sight in his other eye as a result of an infection and became to-
tally blind.

MATHEMATICAL SYMBOLS

PLUS AND MINUS SIGNS

The earliest known printed use of plus (+) and minus (-) signs is in the
book *Mercantile Arithmetic*, which was published by Johannes Widmann
in 1489. He used the signs to indicate surpluses and shortages in cargoes.

Henricus Grammateus began the use of + and - for addition and subtraction in his book *Ayn new Kunstlich Buech* in 1518. The new signs were
not adopted in England until publication of *The Whetstone of Witte*, which
was written in 1577 by the leading English mathematician Robert Recorde
(1510–58). The equal sign (=) was also introduced by Robert Recorde in 1577.

The English rector and mathematician William Oughtred (1575–
1660) was the first to use *x* as a multiplication symbol. He introduced the
x in his great work *Clavis Mathematicae* (*The Key to Mathematics*), which
was published in 1631.

The division sign (÷), known as an obelus, was first used by Johann
Rahn (b. 1622) in his book *Teutsche Algebra*, which was published in 1659.

SLANG

It is thought the use of slang harks back to animism, the first religion. In
animism, it was believed that all objects had an external aspect that
could be perceived by the senses and an invisible aspect that could only
be perceived by a specially gifted person. Thus the use of slang was developed as a way to refer obliquely, rather than directly, to an object (see
"Dictionaries and Encyclopedias," p. 66).

MASSIVE STEPS

COVERING THE MODERN WORLD, THE INDUSTRIAL
REVOLUTION, COMMUNISM, CONSUMERISM, GREAT
INTERNATIONAL COOPERATIONS, AND THE COLD WAR

Don't be afraid to take a big step if one is indicated.
You can't cross a chasm in two small jumps.
—DAVID LLOYD GEORGE (1863–1945)

THE MODERN WORLD

Michael Faraday (1791–1867) is considered by many to be the greatest of all experimental scientists and the originator of modern society.

Faraday was born near London, the son of a blacksmith, and in his youth he was apprenticed to a bookbinder. One of the bookbinder's customers, William Dance, gave Faraday tickets to a series of lectures by the Cornish scientist Sir Humphry Davy (1778–1829), the inventor of, among many other things, the safety lamp for miners. Listening to Davy's lectures stimulated Faraday's interest in scientific experimentation. He quickly applied for a job, but Davy had no position to offer.

Early in 1813 Davy had to dismiss his chemical assistant for fighting in the main lecture hall. This incongruous act had far-reaching consequences for the future of civilization. Faraday heard about the incident, applied for the newly vacant post, and was engaged as Davy's temporary assistant. Shortly afterward Davy embarked on a European lecture tour, and took his new assistant along. During the tour the leading scientists of the day met Davy to exchange views and scientific data. The tour provided an unconventional but unique scientific education for the young Faraday, and on his return to London he soon began to conduct his own experiments.

The experimental work he completed laid the foundations for:

- Radio
- Electric motors
- Electricity generators
- Electrical dynamos
- Electroplating of metals
- The theory of molecular structure
- Understanding the magnetic fields of galaxies
- An early version of the Bunsen burner

Among many other discoveries, Faraday also:

- Developed a type of glass with a high refractive index
- Discovered a number of organic compounds including benzene
- Evolved what was then the radical theory that space was not "nothing," but a medium capable of supporting electrical and magnetic forces
- Discovered that light could be bent by magnetism
- Demonstrated the relationship between electricity and chemical bonding

It is difficult to imagine how different modern society would be without the work of Michael Faraday. Albert Einstein kept a picture of him on the wall of his study, regarding him as having provided the foundation of his own studies.

THE INDUSTRIAL REVOLUTION

The term *Industrial Revolution* refers to, and describes, the process of global change from agricultural and handicraft economies to ones dominated by industry, machine manufacture, and automatic processing. Arnold Toynbee, (1889–1975), the English economic historian, coined the term to describe English economic development in the years 1760–1840.

The area around Coalbrookdale in Shropshire, the site of the world's first iron bridge (completed in 1779), is generally accepted as the cradle of the Industrial Revolution. The area was rich in resources, having ample supplies of coal, iron ore, and limestone within easy reach. The navigable river Severn, which runs through the middle of Shropshire, provided both a convenient means of transport to the docks and abundant water for processing.

The local workforce was only too willing to give up the prevailing seasonal agricultural pay and to take advantage of the higher pay and year-round work offered in the exciting new factories springing up along the banks of the river.

The six main drivers of the Industrial Revolution in Britain were:

- New materials (iron and steel)
- New forms of energy (steam, electricity)
- The invention of new machines (agricultural and industrial)
- The factory system
- Improved communication and transport
- The application of science

To compete with Britain, which had become the dominant force in the world's economy, other European nations underwent the same process of change. Russia and Japan were slow to respond, leaving it until late in the nineteenth century to embrace change. America quickly got into its stride and embraced industrialization with fervor.

The second Industrial Revolution began around 1850 with the advent of the chemical industry, the electrical industry, and the spread of railroads. By the time the second revolution got underway, the world and its civilization had already been changed forever.

COMMUNISM

The founding doctrine of Communism is the liberation of the proletariat: the class within society that lives solely by selling its labor.

In 1847 Friedrich Engels (1820–1925) wrote the draft program for the congress of the League of the Just, which became the Communist League.

The *Communist Manifesto* was written in 1848 by Karl Marx (1818–83) and Friedrich Engels. The manifesto grew out of the perception that poverty was not just an unfortunate consequence of capitalism, but a deliberate creation of the capitalist system. According to the manifesto, capitalism was controlled by the bourgeoisie, and the poverty of the proletariat was seen by Marx as an essential tool in creating wealth. According to Marx, the "labor theory of value" allowed the bourgeoisie to grow massively rich through exploitation of the proletariat's labor.

The revolutionary proposal put forward in the *Communist Manifesto* was that all the means of production and distribution should be owned communally, not by an elite cadre of wealthy businessmen. This looked rather appealing and just, removing, as it did, the need for greed, and distributing commonly held wealth according to the needs of the whole population rather than the wants of a few. Marx proposed that this would eliminate poverty almost at a stroke. Millions listened.

The Bolshevik leader Vladimir Ilyich Lenin (1870–1924) took the idea of Marxist Communism forward with dramatic speed in the Bolshevik Revolutions of 1917, during which he seized power in Russia and created the world's first communist state.

Unlike the capitalist democracies, which had matured over almost two thousand years, within the next half century, Communist regimes—almost all led by dictators—took control of China, Cuba, and North Korea. Major areas of eastern Europe also fell under the influence of Communist Russia as the Soviet empire expanded. At the height of its power, from the

1930s through to the late 1980s, almost half the world's population was governed under some form of Communism. Political commentators everywhere expected further growth.

In the mid-1980s the British and U.S. governments under Margaret Thatcher (b. 1935) and Ronald Reagan (1911–2004) took rigid stances on freedom and economics in opposition to the Soviet Union, which Reagan referred to as the Evil Empire. This began the process that led Mikhail Gorbachev (1931–present), as the last president, to break up the Soviet

Union and restore forms of capitalism and democracy to Russia and its former vassal states during the late 1980s and 1990s.

European Communism's domination ended abruptly between 1989 and 1991, but in 2006, more than 20 percent of the world's population remains under communist control in China, North Korea, Cuba, Laos, and Vietnam.

CONSUMERISM

The consumer is not a moron, she's your wife.
—DAVID OGILVY (1911–99)

The groundbreaking book, *Unsafe at Any Speed: The Designed-In Dangers of the American Automobile*, was written by a young American lawyer, Ralph Nader (b. 1934), and published in 1965 to massive public acclaim.

Nader claimed in his book that in the 1950s and early 1960s hundreds of occupants of motorcars were dying every year in accidents that could have been prevented. Better design (not to be confused with styling) and the installation of safety features were the changes he advocated.

Nader asserted that, rather than coming to agreement between themselves on universal standards of safety, the big car manufacturers were continually wasting their time and efforts by lobbying the U.S. government to prevent safety measures being introduced on a compulsory basis.

Resulting from the impact his book had in the United States and Europe, among other things the fitting and wearing of seat belts was made compulsory. Almost as importantly, the capacity of tires and suspension to support a fully loaded vehicle had to be proven. Crash testing was introduced (even on Rolls-Royces), and air bags were eventually fitted as standard.

Nader's campaigns affected most industries. Consumers began to realize they had powerful rights, and they demanded improved safety and better information. Governments on both sides of the Atlantic responded with stricter controls. Among other things, the individual contents of prepared foodstuffs were compulsorily listed, to show the amounts of additives.

Before the publication of Nader's book, consumers had a tendency to trust big corporations and, at least subliminally, to associate big business with safe products and government with being trustworthy. Consumers' rights under the law led to landmark court cases against companies that foisted unsafe products onto the general public. It was no longer enough for a producer to claim caveat emptor (let the buyer beware). Once the genie of corporate responsibility was out of the bottle, continuous improvements were instituted in aviation, food, air pollution, water pollution, forestry, and the behavior of lawyers, among others.

Nader also campaigned against mass advertising, on the premise that it creates false desires in consumers, and he argued strongly against sensationalism in the daily news.

GREAT INTERNATIONAL COOPERATIONS

You cannot put a rope round the neck of an idea.
You cannot confine it in the strongest prison cell.
—SEAN O'CASEY (1880-1964)

Occasionally, in the face of technological advances and the threat of public pressure, nations cooperate to make life better and improve conditions for everyone, not just their own voters. Nationalism and intolerance are put aside to allow great advances to take place.

INTERNATIONAL TELEGRAPHIC UNION

The International Telegraphic Union (ITU) was set up in 1865, originally with twenty member countries, to coordinate all international telegraphic traffic. The union was the initiative of the French government. It was authorized and ratified at the 1865 Paris Convention (Convention Télégraphique Internationale de Paris) to: regulate radio frequencies, assist with technical and operational matters, and promote telegraphic development in countries not yet "telegraphed."

Britain was originally excluded from participation in the ITU, finally being admitted in 1871.

Despite the first telegraphic message being sent in 1844, until the ITU Agreement in 1865 there had been no common international telegraphic traffic standard. Since each country used a different system, messages had to be transcribed and handed over at the border. The messages then had to be translated and retransmitted using the neighboring country's telegraphic system.

The organization has been renamed the International Telecommunication Union.

UNIVERSAL POSTAL UNION

The Universal Postal Union (UPU) was set up in 1874 under its original name, General Postal Union. The name was changed in 1878.

The objective of the union was to organize and coordinate postal services throughout the world. The most important agreement was that postage paid at source in one country would cover the cost of delivery in another. It was also agreed that all national stamps would be accepted in all other countries.

World Post Day is held on October 9 each year, the anniversary of the establishment of the UPU.

RED CROSS AND RED CRESCENT

The Red Cross and Red Crescent were the brainchild of Swiss businessman and humanitarian Jean-Henry Dunant (1828–1910). The decisive moment came for Dunant in the aftermath of the 1859 Battle of Solferino in the Franco-Austrian War. He witnessed helpless, wounded soldiers lying on the battlefield, being systematically bayoneted or shot to death.

Dunant had been involved in the organization of emergency aid for both French and Austrian wounded and had been sickened by the suffering on both sides. In a single day, the total killed was over five thousand and the total wounded was over 23,000. Dunant was strongly affected by the mistreatment of the wounded and determined to make changes.

In 1859 he proposed the formation of societies for the relief of suffering, in all countries, and was awarded the first Nobel Peace Prize in 1901.

The Geneva Convention of 1864 committed all the signatory governments to care for all the wounded, whether enemy or friend. From 1906, at the insistence of the countries of the Hapsburg Empire, the Red Cross became the coordinating body in Christian countries and the Red Crescent in Muslim countries. The cross was regarded as a potent symbol of Christianity and unsuitable for predominantly Muslim countries.

HAGUE TRIBUNAL

The Hague Tribunal is the popular name for the Permanent Court of Arbitration (PCA) based at The Hague in Holland. The PCA was established in 1899 by a convention of the First Hague Conference, as a means to resolve international disputes. The initiative came from Czar Nicholas II of Russia (1868–1918).

Currently, 104 states are party to the treaty.

LEAGUE OF NATIONS

Following the initiative of the British Foreign Secretary Edward Grey (1862–1933), the League of Nations was proposed by U.S. President Woodrow Wilson (1856–1924) at the Paris Peace Conference of 1919.

Wilson's objective was to create an organization with enough strength to prevent future conflicts like the First World War. After Wilson returned home from Paris, the U.S. Senate refused to ratify U.S. membership in the League of Nations.

After failing to prevent the Second World War, the League of Nations dissolved itself and transferred all its services and property to the United Nations.

UNITED NATIONS

The United Nations (UN) was formed with the intention of "encouraging international cooperation in solving economic, social, cultural, and humanitarian problems." The name United Nations was formally adopted on October 24, 1945, to mean the organization as it is recognized today, an association of peace-loving nations that accept the obligations of the UN charter. The original Big Four were the United States, the Soviet Union, the United Kingdom, and China, with France added later and offered a permanent seat on the Security Council.

The term *United Nations* originally referred to the Allies in the Second World War, who had agreed on their wartime alliance in 1942. The agreement prevented any individual nation seeking a separate peace treaty with any of the Axis powers (Japan, Germany, and Italy). The three war leaders, Roosevelt, Stalin, and Churchill, agreed on the original charter, which included objectives for humanitarian aid and human rights as well as the organization's peacekeeping role.

UNIVERSAL DECLARATION OF HUMAN RIGHTS

The Universal Declaration of Human Rights was completed by the United Nations Commission on Human Rights in June 1948 and was adopted by the General Assembly in December of the same year.

Eleanor Roosevelt (1884–1962), the widow of U.S. president Franklin D. Roosevelt (1882–1945), and chairwoman of the committee that drafted the declaration, described it as the "Universal Magna Carta for all mankind." The declaration was universally adopted, with the exception of six abstentions by the Soviet bloc plus Saudi Arabia and South Africa.

NORTH ATLANTIC TREATY ORGANIZATION

After the signing of the North Atlantic Treaty in 1949, the North Atlantic Treaty Organization (NATO) was formed to provide a counterweight to the growing and perceived threat of aggression from the Soviet Union. The United States and Canada formed a military alliance with ten European countries: Belgium, Denmark, France, Iceland, Italy, Luxembourg, the Netherlands, Norway, Portugal, and the United Kingdom.

NATO was instrumental in maintaining peace in Europe for more than forty years, not only from the Soviet threat, but also between other previously warring European nations.

THE METRIC SYSTEM

The metric system was developed in France during the eighteenth century to provide a uniform system of measurement to replace the widely differing systems then in existence. Metric measurement includes weight, volume, length, area, capacity, and temperature and is based on multiples being to the power of ten. The intention is that measurements can be understood throughout the world and universal standards adopted in places such as research laboratories.

Frenchman Charles Maurice de Talleyrand-Périgord, Prince de Bénévent, Bishop of Autun (1754–1838), known more popularly as Talleyrand, and Sir John Riggs Miller (d. 1798) of England, jointly championed the metric system in the 1790s.

By 1840 France made metrication mandatory. The metric system was made legal in the United States in 1866, but not mandatory. At the Metric Conference of 1875 in France, seventeen additional countries signed the Treaty of the Meter. Britain signed in 1884, to make the metric system a common international measurement.

The postmaster general, Tony Benn (b. 1925), headed the British Metrication Board set up in 1969. The metric system was finally adopted in Britain in 1970, and the currency was decimalized in 1971, although to this day there remain many anomalies. The remaining imperial measurements include miles, pints (of beer), and acres.

ORGANIZATION OF PETROLEUM EXPORTING COUNTRIES

A fair price for oil is anything
you can get plus 10 to 20 percent.
—ANONYMOUS

Established in 1960, the Organization of Petroleum Exporting Countries (OPEC) was formally constituted in January 1961 by Iran, Iraq, Kuwait, Saudi Arabia, and Venezuela. Its stated aim is to coordinate the petroleum policies of its members. Effectively, OPEC is a cartel set up to con-

trol the volume of oil produced by each of its member states. By controlling the volume, OPEC also controls the price of oil, and therefore controls a major part of the world economy. OPEC countries hold around 70
percent of the world's oil reserves.

THE COLD WAR

The cold war can be dated from 1947, and was so called because no direct, armed conflict took place. The cold war was brought about after the
end of the Second World War by the mutual suspicions of the United
States and its allies on the one side, and the Soviet Union and its allies
on the other.

It was a war that was also, at times, fought by proxy. Armed conflicts
in Vietnam and Angola were such proxy wars between the two opposing
cultures of the West and East.

The leaders of the United States and Soviet Union officially agreed the
end of the cold war in 1989.

HOTLINE

In 1963, at the height of the cold war, the first hotline was installed
between the White House in Washington DC and the Kremlin in
Moscow (where it was known as the "red telephone"). It was commissioned after the Cuban missile crisis of 1961, during which the world had
watched with bated breath as the two superpowers faced each other with
nuclear weapons. The leaders of the United States and the Soviet Union
agreed that they should find a means to speak directly and immediately
to each other. The line was made available, with no interference other
than by the respective interpreters, at any time of day. Up to that point,
there had been a real danger of accidental nuclear war being triggered
by deputies while the leaders were unable to contact each other.

The original apparatus used for the hotline was an old teleprinter (an
electromechanical device, similar in appearance to a typewriter, for transmitting the written word between two points via simple electrical wires).
A real telephone was not installed until 1970.

Although it was a cooperation between only two nations, the hotline had worldwide safety implications.

THE WARSAW TREATY OF FRIENDSHIP, COOPERATION, AND MUTUAL ASSISTANCE

The Warsaw Treaty was signed in May 1955 to form an alliance between Eastern bloc countries, led by the Soviet Union and known as the Warsaw Pact. The treaty was signed to counter the perceived military threat from the NATO alliance, and the perception became more acute once West Germany had been admitted to membership in NATO in 1952.

The participating countries of the Warsaw Pact were the Soviet Union, Albania, Bulgaria, Czechoslovakia, East Germany, Hungary, Poland, and Romania.

In 1989 the Warsaw Pact became redundant and it was declared nonexistent in 1991.

QUESTIONABLE ORIGINS

COVERING PRESIDENT OF THE UNITED STATES,
FOSBURY FLOP, MODERN OLYMPICS,
INTERNAL COMBUSTION ENGINE, PNEUMATIC TIRE,
LIGHTBULB, DISCOVERY OF AMERICA, RADIO,
DNA, SPEAKING MACHINE, LSD,
THE FIRST POWERED FLIGHT, AND BOND GIRLS

Throughout history people have been incorrectly attributed with achievements, new inventions, and discoveries, as this chapter explains.

PRESIDENT OF THE UNITED STATES

The first president of the United States is generally acknowledged to be George Washington (1732-99), who served as president from 1789 to 1797. And yet, in 1781, while Washington, as commander in chief of the Continental army, was still leading the war against Britain, John Hanson (1715-83) was elected as the first "President of the United States in Congress Assembled." It would be eight more years, and six more presidents, before the presidency would go to George Washington.

In 1776, America had declared its independence from Britain, and in the same year Congress proposed the Articles of Confederation. These articles were intended to bind together the loose confederation of independent states, but, because of territorial disputes between some of the states, the articles were not signed until 1781.

John Hanson was instrumental in bringing the last state, Maryland, his home state, into the confederation. The new Articles of Confederation stipulated that the United States should appoint a president each year, and that the person appointed should hold the post for the maximum period of one year in any three. In the year of the signing, 1781, Hanson was unanimously voted in as the first president and served for one year, the period laid down in the articles.

During his term of office, Hanson established the great seal of the United States, the post of secretary of war, the treasury department, and the foreign affairs department, and he ordered all foreign troops off American soil. The six presidents who followed each had a year in power, until Congress replaced the Articles of Confederation with the American Constitution in 1789.

Despite carrying the title of president, under the Articles of Confederation, the office was not precisely that of chief executive of the country, but was more closely related to that of Speaker of the House.

In 1789 George Washington was unanimously elected as the first president under the newly written Constitution and served until 1797.

FOSBURY FLOP

Dick Fosbury (b. 1947) of the United States is wrongly credited with being the first to use the then-revolutionary style of high jumping that bears his name. There is no dispute that Fosbury won the 1968 men's Olympic gold medal for the high jump using the "Fosbury flop," which no other jumper of the time used. Nor is it disputed that he had independently developed the technique, but he was not the original.

The first person to use the "flop" method—launching himself at the high-jump bar with an arched back and facing skyward—was Bruce Quande, also an American, in 1963, and there are photographs to prove it.

MODERN OLYMPICS

What are termed the modern Olympics took place for the first time in Athens in 1896.

Nearly half a century earlier, in 1850, Dr. William Penny Brookes (1809–95) staged the first Olympian Games in Much Wenlock, a small Shropshire town. The games were held annually, and by 1865 the number of spectators at the event had increased to ten thousand, with competitors making their way to the games from all over Europe. Dr. Penny Brookes tried to generate interest from the Greek government to stage the same track and field events held at Much Wenlock, in Athens. Unfortunately he met with no success.

But then, in 1856, Evangelis Zappas, a wealthy Greek businessman living in Romania, wrote to King Otto of Greece offering to fund a revival of the Olympic Games. Zappas proposed that the new games should be held for all time in Greece. His sponsorship was accepted, and the first of the Zappa-sponsored Olympic Games was staged in Athens. Dr. Penny Brookes donated prize money for one of the events. Zappas financed a second games in 1865 and a third in 1870. The third games were for elite competitors only and were the last of the Zappas-funded games.

Baron Pierre de Coubertin (1863–1937), the acknowledged founder of the modern Olympics, visited Much Wenlock in 1890 at the invitation of Dr. Penny Brookes, and a special games was put on in his honor. The baron wrote an article in the French sporting magazine *La Revue Athlétique*, in which he said, "If the Olympic Games are revived, it will not be due to a Greek but to the efforts of Dr. W. P. Brookes."

The modern Olympic movement Inspired by his visit to Much Wenlock, six years later Baron de Coubertin went on to found the modern Olympic movement, and to stage the first of the modern Olympic Games

in 1896 in Athens. Sadly, William Penny Brookes died a few months before the games in Athens, unacknowledged and almost unheard of, except in Much Wenlock. The annual Olympian games continue in Much
Wenlock, dedicated, as is a local museum, to the memory of Dr. William
Penny Brookes.

It is interesting that over two hundred years earlier, Robert Dover's
"Olimpick Games" were being held annually in the Cotswold town of
Chipping Campden. The first of these Olimpick games was held in 1612,
and the games were held annually until 1852. There was a gap of a century until they were revived in 1951 for the Festival of Britain. Despite its
eminent title, the Olimpick Games was never more than a large village
sporting festival, with shin-kicking as one of the events.

INTERNAL COMBUSTION ENGINE

In 1807 Joseph Nicéphore Niépce (see "Inventions," p. 251) was granted
a patent for a pyréolophore, powered by coal and resin. In 1826 Samuel
Morey (1762–1843), a prolific inventor who lived in the American state
Vermont, was granted a patent on an internal combustion engine. Morey
wrote a paper on his invention in the *American Journal of Science and
Arts*, in the year of the granting of the patent, entitled "An Account of a
new Explosive Engine."

Morey's invention predates Lenoir's generally acknowledged first successful internal combustion engine by thirty-four years and that of Otto
by thirty-eight years.

PNEUMATIC TIRE

John Dunlop (1840–1921) is recorded as the inventor of the pneumatic
(air-filled) tire, in 1888. However, the pneumatic tire was actually invented by a twenty-three-year-old Scot, Robert Thomson (1822–73), who

applied for a patent in 1845 for what he called an "aerial wheel." Dunlop had merely improved the original design, at the request of his son, by filling the tire with air to help make riding his bicycle more comfortable.

LIGHTBULB

The great American inventor Thomas Alva Edison (1847–1931), along with his thousand other patents, has received almost all the credit for inventing the electric lightbulb in 1879.

However, the lightbulb had actually been around since 1809, and was invented by Sir Humphry Davy (1778–1829), in England. Davy connected two wires to a battery and attached a charcoal strip between the other ends of the wires. The charcoal glowed, making the first arc lamp.

In 1875 Henry Woodward and Matthew Evans patented a new version of the lightbulb in Toronto. Unable to raise the capital to produce their lamps on a commercial scale, they sold out to Edison, who went on to improve the design and use a less powerful electric current.

In 1879 Edison in America, and Joseph (later Sir Joseph) Swan (1828–1914) in England, simultaneously found a way to make the lightbulb a commercial proposition by giving it longer lasting qualities.

Heinrich Göbel (1818–93), an American of German descent, produced an incandescent lamp in 1854, twenty-five years earlier than Edison and Swan, using carbonized bamboo as a filament. By 1859 Göbel had improved his lamp so that it would last up to four hundred hours. He took out a court case against Thomas Edison in 1893, claiming to be the inventor of the electric lightbulb. The court accepted Göbel's claim and recognized him, albeit wrongly, as the inventor of the electric lightbulb. A few months after the court's decision, Göbel died of pneumonia.

DISCOVERY OF AMERICA

Was America discovered by Christopher Columbus (1451–1506) in 1492, as is almost universally taught? Or was it discovered in 1497 by the Italian explorer John Cabot (circa 1425–1500) on an expedition sponsored by the wealthy British aristocrat Richard Amerike? Or was it perhaps discovered in the tenth century by Vikings?

And now, to make matters even more confusing, there are Chinese claims that Admiral Zheng He went to America in 1421.

On his voyage west in 1492 Columbus was not looking to discover new lands. He had been brought up as a militant Christian, and was looking for a way to attack Muslims unexpectedly from the rear. The man who actually spotted land, most likely the island of San Salvador in the Caribbean, for the first time was Martin Alonzo Pinzon.

Richard Amerike sponsored John Cabot and asked him to name any "newfound lands" after himself. There is a strong possibility that Cabot sailed to the east coast of America and named it America after Amerike.

Admiral He is supposed to have circumnavigated the earth, beginning his journey in 1405, and finally returning to China, not only having discovered America, but also Australia and Africa. This massive journey of 31,000 miles is depicted on an ancient chart, which is now displayed in Beijing.

The Viking explorer Lief Ericsson (circa AD 980–1025) has a strong claim to have been the first European to set foot on the east coast of America after a sea journey westward from Scandinavia around the year AD 1000. His discovery is recorded in *The Saga of Eric the Red*, a book written in 1387 by Jon Thordharson (Eric was Lief's father).

The original occupiers of the land of North America came from Asia, migrating across the land bridge that had linked Alaska with Russia 20,000 years earlier. DNA tests and language comparisons have been conclusive in establishing the link between the Native American population and East Asians (see "Countries and Empires," p. 36).

RADIO

Guglielmo Marconi (1874–1937) is widely thought to have invented radio in 1896, and is known as the father of radio. He did indeed pioneer the first long-distance radio broadcasts, and managed to have a patent granted, but Marconi did not invent radio itself.

In 1943, the U.S. Supreme Court overturned Marconi's patent in favor of the Serbian inventor Nikola Tesla (1856–1943), who has now been properly credited as radio's inventor. Tesla died a few months before the final vindication of his work.

DNA

On February 28, 1953, James Watson (b. 1928) and Francis Crick (1916–2004) announced that they had discovered the famous double helical shape of the DNA molecule. Together with Maurice Wilkins (1916–2004), they were awarded the 1962 Nobel Prize for Physiology or Medicine. DNA had actually been identified in 1869 by Swiss scientist Johann Friedrich Miescher (1844–95) working in Tübingen, Germany, but the actual structure had always remained elusive (see "Health," p. 130).

At a competing laboratory, at King's College, London, Wilkins had been working with a brilliant young scientist, Rosalind Franklin (1920–58), whose pioneering use of X-ray crystallography to look into the structure of molecules was beginning to yield results. Relations between Wilkins and Franklin were not particularly good, and in an act of savage betrayal Wilkins revealed Franklin's research results, and the famous Photo 51, to Watson and Crick. They quickly seized on this windfall and published their own paper, without acknowledging Franklin's work. It is widely thought that the wrong people were awarded the Nobel Prize.

Rosalind Franklin's work has been properly recognized only in recent years. It is one of the tragedies of medical history that she did not survive long enough to reap the reward for her work, dying at only thirty-eight, of ovarian cancer. The cancer was almost certainly brought on by her heavy exposure to X-rays (see "Health," p. 129).

SPEAKING MACHINE

In 1788 Johann Wolfgang Ritter von Kempelen de Pázmánd (1734–1804) invented the first "speaking machine." The device used a set of bellows to pump air across a reed, which in turn excited a hand-varied resonator to produce the sound of a voice. The human voice was reproduced on the Kempelen machine, but not recorded. De Pazmand's device trumped Edison's phonograph by a century.

LSD

Generally thought of as a 1960s hallucinogenic drug, LSD (lysergic acid diethylamide) was actually synthesised in 1938 by the Swiss chemist Dr. Albert Hofmann (b. 1906). When he repeated the experiment in 1943 and accidentally licked his fingertips, he discovered the hallucinogenic properties of LSD.

Hofmann insisted on continuing his experimental work by trying the drug out on himself. Not knowing how much LSD to take, he accidentally took three times what would come to be regarded as the normal dose, and ended up on a massive hallucinating "trip."

On January 11, 2006, Dr. Hofmann reached his one hundredth birthday. To celebrate the occasion, a three-day international psychedelic conference was held in Basel, Switzerland, on January 13–15, 2006. There were more than two thousand delegates from thirty-seven countries, and it was attended by writers, artists, and scientists as well as friends and the press.

USA Today reported Dr. Hofmann saying, "I produced the substance as a medicine. It's not my fault if people abused it." He also claimed, "I had wonderful visions."

LSD was marketed by Sandoz Laboratories in 1947 as a cure for schizophrenia and became freely available. The first mass-produced LSD specifically for recreational use came from the laboratory of the renegade chemist Augustus Owsley Stanley III in 1965. In 1967 he was sentenced to three years for possession of a huge quantity of the drug—the equivalent of 100,000 doses.

THE FIRST POWERED FLIGHT

The Wright brothers, Orville and Wilbur, flew their powered airplane, the Flyer, for the first time at Kitty Hawk in North Carolina on December 17, 1903, and entered the history books as the pioneers of powered flight.

Yet, in New Zealand on March 31, 1903, farmer Richard Pearse, known as Bamboo Dick, flew 500 feet in a powered airplane, landed in a gorse hedge, and wrecked his machine. He had beaten the Wright brothers by eight months, but Pearse was a man of high standards and refused to claim his rightful place in the history books. He felt that the Wright brothers had landed under control, whereas he had not. However, that does not alter the fact that Richard Pearse was the first human being to achieve flight in a heavier-than-air aircraft under power.

BOND GIRLS

The immortal image of Swiss sex symbol Ursula Andress (b. 1936) emerging from the Caribbean in a white bikini in the first of the James Bond films, *Dr. No*, has become an icon of the cinema. Most filmgoers would opt for Miss Andress, who was playing Honey Ryder, as the original Bond girl.

However, Ursula Andress was beaten to this honor by British actress Eunice Gayson (b. 1931), who appeared in the opening sequence of *Dr. No*, playing Sylvia Trench. According to Miss Gayson, she was instructed to take Sean Connery out for a drink to help him relax because he kept fluffing his famous line, "The name's Bond. James Bond." He accidentally kept giving his own name.

> *If the role demands it, then*
> *naturally I will remove my clothes.*
> —URSULA ANDRESS (B. 1936)

RELIGIONS

COVERING ANIMISM, SUN WORSHIP, MOON WORSHIP,
JUDAISM, HINDUISM, ZOROASTRIANISM,
CONFUCIANISM, DRUIDISM, CHRISTIANITY, BUDDHISM,
ISLAM, SHINTO, VOODOO, MORMONISM, SCIENTOLOGY,
MOONIES, AND ATHEISM

If God did not exist, it would be necessary to invent him.
—VOLTAIRE (1694–1778)

Not only is there no God,
but try getting a plumber on weekends.
—WOODY ALLEN (B. 1935)

ANIMISM

The oldest belief system of all, animism predates all known organized religions. The word comes from the Latin *anima,* meaning "breath" or "soul." It originated in the Upper Paleolithic Period between 40,000 and 10,000 BC.

The core belief of animism is that spiritual beings exist separately from the physical being. In animism, the breath is the soul. When breathing stops, the soul passes to another life. It holds that the soul passes into plants, animals, and even inanimate objects after the death of the person. This is thought by some experts to have been a way for primitive peoples to explain sleep, dreams, and death.

SUN WORSHIP

Agrarian societies whose crops and livelihood were dependent on the weather mainly practiced sun worship. It reached its zenith in ancient Egypt around 2500 BC, with one of the most important gods being Ra, the sun god. The pharaoh was regarded as the son of Ra and his representative on earth.

Other societies had different names for their sun god. In Mesopotamia the sun god was called Shamash, and the ancient Greeks had two sun gods, Apollo and Helios. The Incas, Aztecs, druids, and Zoroastrians all worshipped the sun.

MOON WORSHIP

The Sumerians of around 2500 BC, and other ancient worshippers of the moon, held ceremonies on the last day of the year, when the sun, who was king for a year, was sacrificed to the moon. Ceremonially, the sun's genitals are removed, the blood is drunk, and the testicles are eaten. Then, during the ceremony, the moon resurrects the sun so that life may continue on earth.

Moon worship is not to be confused with the cult of the Moonies, which is a twentieth-century phenomenon.

JUDAISM

One of the most ancient belief systems, Judaism is the religion of the Jews. It was founded in around 1900 BC by Abraham, in Canaan, which is roughly equivalent to modern Israel and Lebanon.

Judaism is the oldest of the monotheistic faiths, which hold that there is only one God rather than a plethora of many different deities. The name of the Jewish God is Yahweh (Jehovah), and the holy book is the Torah.

Circumcision The Hebrew holy book, the Torah, which according to some was handed down to Moses in 1280 BC, refers to Jehovah's command to Abraham to circumcise himself, his sons, his slaves, and servants.

The origin of the practice of circumcision is uncertain, but its roots can be traced back to around 6000 BC and the puberty rites of tribes in northeastern Africa and the Arabian Peninsula. From 3100 BC the ancient Egyptians circumcised boys between six and twelve years old as a mark of ritual cleansing.

In AD 570 the prophet Mohammed was born supposedly already circumcised. This inspired Muslims to circumcise boys shortly after birth.

HINDUISM

Religious scholars believe it is conceivable that the roots of Hinduism in the Indus Valley in India may stretch as far back as 4000 BC, which would make it the world's oldest organized religion. However, alternative modern theories consider it is more likely that Hinduism was founded around 1500 BC.

Hindus are polytheistic. They believe in many gods, and unlike in Christianity, Islam, and Judaism, there is no single, named god to worship. Hinduism accepts the view that there is only one reality, but all formulations of the truth are respected. Many Hindu representations of the Divine Being are female, such as Durga, the protective mother, and Lakshmi, the goddess of prosperity, purity, chastity, and generosity. Hinduism holds that without honoring female qualities, religion is incomplete.

The holy books of Hinduism are the Vedas, and the key to Hindu belief is the transmigration of the soul, which produces a continuous cycle of birth, life, death, and rebirth. The individual soul enters a new existence once the body has died, and the total of all previous moral conduct will determine the quality of the soul's rebirth.

ZOROASTRIANISM

The ancient religion of Persia (modern Iran), Zoroastrianism was founded by the prophet Zarathustra (also known as Zoroaster). There is a great deal of debate among academics about when the religion was founded, ranging from between the eighteenth and eleventh centuries BC to approximately 600 BC.

Zoroastrianism is known as "the threefold path." Its motto, as written in the holy book, the Avesta, is "Good thoughts; Good words; Good deeds." Zarathustra preached that there is only one God, Ahura Mazda (Wise Lord). The basic scripture is a set of five poetic songs called the Gathas, composed by Zarathustra, each with seven attributes:

1. Good thought *(animals)*
2. Justice and truth *(fire and energy)*
3. Dominion *(metals)*
4. Devotion and serenity *(earth)*
5. Wholeness *(water)*
6. Immortality *(plants)*
7. Creative energy *(humans)*

Zoroastrianism survived invasions of its Persian homeland by Greeks, Muslims, and Mongols, and is still practiced in India by the Parsis. In the eighth century, the leader of the Parsis made representations to Jadav Rana, the king of Sanjan (modern Gujarat in India) to be allowed to bring his people into India to escape persecution in Persia. The king was reluctant to agree to this request, since India was already becoming over-crowded. He filled a cup to the brim with milk to demonstrate that there was no more room. The leader of the Parsis took a spoonful of sugar, dropped it into the cup of milk, and stirred it, saying, "We will be like the sugar in this milk. You will not be able to see us, but we will help to sweeten your lives." The king accepted the Parsis, with the condition that they adopted local dress and did not proselytize their religion.

CONFUCIANISM

Man has three ways of behaving wisely.
First by meditation; that is the noblest.
Second by imitation; that is the easiest.
Third by experience; that is the bitterest.
—CONFUCIUS (551–479 BC)

K'ung-Fu-tzu (551–479 BC), known in the West as Confucius, founded Confucianism in China in the fifth century BC, and it was adopted as the state religion of imperial China.

The holy books of Confucianism, the Si Shu Wu Jing (Four Books and Five Classics), teach the practical values of benevolence, reciprocity, respect, and personal effort, making Confucianism more an ethical and philosophical system than a formal system of religion with a named god.

DRUIDISM

The first records of druidism appear around 200 BC, and Julius Caesar (100–44 BC), gives an account of this religion in his *History of the Wars in Gaul*, written between 59 and 51 BC. Both the ancient Greeks and Romans believed that Celtic magicians, as druid leaders were thought to be, held special knowledge and the ability to channel their power for good or evil through gods, spirits, and ancestors. The Roman author and naturalist

Pliny the Elder (AD 23–79), who died in the eruption of Vesuvius that buried Pompeii, wrote of druids being learned men and women having "oak knowledge," from the ancient words *dru* (oak) and *wid* (know).

There are druids today and one of the most famous gatherings is the annual Welsh eisteddfod. They also believe in the transmigration of souls and that they have all descended from a common ancestor.

CHRISTIANITY

Christianity is one of the major monotheistic religions of the world, with a belief in a single named God.

Jesus Christ was born in Bethlehem around 4 BC, and it is believed by followers of Christianity that he was the son of God. Jesus used parables and sermons to make his messages memorable and easily understood, and preached love and forgiveness and tolerance. He began teaching publicly at the age of thirty, after he had been baptized in the river Jordan by John the Baptist, and continued for the last three years of his life.

After Jesus' death around AD 30, his authority passed directly to the apostle Peter (later Saint Peter), who died in either AD 64 or 69. Peter is regarded by the Roman Catholic Church as the first pope. The Gospels, the books recounting the life and teachings of Jesus, were not written until around AD 100, seventy years after his death.

The first Christian church is considered to be Saint Mark's house in Jerusalem, the place Jesus chose for the apostles to eat Passover before his crucifixion. The Holy Spirit is said to have descended into Saint Mark's house on the day of Pentecost.

Saint Mark was martyred in AD 68. He was tied to a horse's tail and dragged through the streets of Bokalia in Alexandria for two days. In keeping with the traditions of the day, parts of his body were preserved. His head is now in a church in Alexandria, other parts of him are in Cairo, and the remainder is in St. Mark's cathedral in Venice.

The Roman emperor Constantine (AD 272–337)

converted to Christianity in AD 313. His conversion led to the rise of Rome as the center of the Christian faith and the church's influence over the world, which lasts to this day.

Christmas Day was first recognized and celebrated as December 25 in the fourth century.

The first king in England to convert to Christianity was King Aethelbert of Kent. In AD 597 Augustine (later Saint Augustine) landed in Kent on a mission from Pope Gregory to convert the pagan Aethelbert and spread the faith throughout the country. Aethelbert, whose queen Bertha was already a Christian, converted most probably in AD 597 and allowed Augustine and his followers to stay in an old church in Canterbury. Augustine became the first Archbishop of Canterbury and worked with Aethelbert to draw up the first Anglo-Saxon laws.

The Great Schism of AD 1054 divided Christianity into the Catholic (Western) and Orthodox (Eastern) Churches. The pope, as head of the Western Church, and the patriarch of Constantinople, as head of the Eastern Church, excommunicated each other in a dispute over whose authority was the greater.

The Coptic Orthodox Church developed as the main Christian church in Muslim Egypt in the seventh century. *Copt* is from the Arabic *gibt*, which itself is a corruption of *Aigyptios* (Egyptian).

The first Christmas cards originated in 1840. The letter *X* in the word *Xmas* is not used just to save writing out the whole of the word *Christmas. X* is *chi*, the first letter of the Greek word *christos*, meaning "anointed."

ROMAN CATHOLICISM

The head of the Roman Catholic Church is the pope (a title that derives from the Latin word for "father"). Since the ninth century the pope has also been Bishop of Rome. The full title of the pope is: Bishop of Rome, Vicar of Jesus Christ, Successor to the Prince of Apostles, Supreme Pontiff of the Universal Church, Patriarch of the West, Primate of Italy, Archbishop and Metropolitan of the Roman Province, Sovereign of the State of the Vatican City, Servant of the Servants of God.

The first pope was Saint Peter, the leader of the apostles and therefore of Christianity until his death in AD 64 or 69.

New popes are appointed by a process called the conclave. This is a form of election that began in 1274 and starts fifteen days after the death of the previous pope. The fifteen-days rule was introduced to allow cardinals time to travel to Rome from all corners of the Roman Catholic world.

Society of Jesus (Jesuits)

The Jesuits were formed in Paris in August 1533 by Ignatius of Loyola, later Saint Ignatius (1491–1556), a Spanish soldier and nobleman. Ignatius and six other young men, who had met at the University of Paris, formed the first group. They drew up laws that demanded chastity, poverty, and pilgrimage to Jerusalem.

On August 27, 1540, Pope Paul III gave his approval to the new order, and by the time of Ignatius's death in 1556, one thousand Jesuits had been recruited.

PROTESTANTISM

The Reformation of the sixteenth century—the movement to protest against the absolute rule of the Catholic Church of Rome—saw the establishment of Protestantism.

The most dramatic single act of the Reformation was the nailing of ninety-five theses to the door of Wittenberg church by the German theologian Martin Luther (1483–1546) in 1517. The theses condemned greed and worldliness in the Catholic Church.

Luther was an Augustine monk whose teachings inspired the Lutheran and Protestant movements. He married in 1525, setting the example for clerical marriage.

The Reformation led ultimately to the split of Protestantism from the Roman Catholic Church.

Church of England

Christianity in England can be traced to the second century, existing independently of the Roman Church. Church records show that English bishops attended the Council of Arles in southern France, which was held in AD 314, more than two hundred years before Saint Augustine (d. 604) was sent by Pope Gregory to convert England to Christianity in AD 597.

The Church of England as we know it, which is the mother church of the worldwide Anglican Union, has its roots in the reign of King

Henry VIII in the sixteenth century. Henry broke away from the Roman Catholic Church when Pope Clement VII refused to approve the annulment of his marriage to Catherine of Aragon. An act of Parliament of 1534 made Henry (and all future British monarchs) the head of the Church of England, and he subsequently refused to recognize the authority of the pope in England. Another act of Parliament of 1534 recognized the annulment of his marriage to Catherine of Aragon and his marriage to Anne Boleyn.

GREEK ORTHODOX

The Greek Orthodox Church is a recent addition to the world's Christian religions, and presently has 10 million adherents. After the Greek War of Independence (1821–32), negotiations were begun with the patriarch of the Eastern Orthodox communion for an independent Greek church. The Church was finally recognized in 1850 and is organized along the lines established by the Russian emperor Peter the Great for the Russian Orthodox Church.

BUDDHISM

A religion of philosophy, Buddhism follows the teachings of Siddhartha Guatama (Buddha) who, it was previously thought, lived from 566 BC to 486 BC. This has recently been revised to 490–10 BC.

Ashoka the Great ruled the Mauryan empire, which covered a major part of Asia, from 273 to 232 BC. He converted to Buddhism and gave it his royal patronage, which led to the rapid expansion of the Buddhist faith throughout Asia. The lack of a written gospel from the Buddha helped lead to the formation of more than four hundred diverse movements within Buddhism.

The Buddhist holy book, known as the Tripitaka, was not written until more than four hundred years after the death of the Buddha. Until then, the method used to preserve his teachings was for student monks to learn them by heart. At group councils of the monastic orders, a senior monk would ask a question, and the assembled monks would recite the appropriate teaching. It was the Theravadian monks who wrote down the Tripitaka in the first century. The teaching of Buddhism is based on the four Noble Truths, which lead to the eightfold path:

1. Life is disappointing
2. Suffering comes from the desire for pleasure, power, and continual existence
3. To stop suffering—stop desiring
4. Stop desiring by following the eightfold path, being the rightness of:

- Views
- Intention
- Speech
- Action
- Livelihood
- Effort
- Awareness
- Concentration

ISLAM

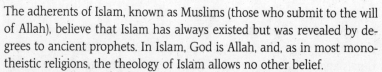

The adherents of Islam, known as Muslims (those who submit to the will of Allah), believe that Islam has always existed but was revealed by degrees to ancient prophets. In Islam, God is Allah, and, as in most monotheistic religions, the theology of Islam allows no other belief.

In AD 610 the final revelation was made to the Prophet, Mohammed (circa AD 570–632), as he meditated alone in a cave. According to tradition, the Angel Jibreel (Gabriel) visited Mohammed and instructed him to recite the word of Allah. During the rest of his life he continued to receive revelations, which were written down as the Qu'ran (Koran).

The other holy book of Islam is the hadith, which is an account of the spoken traditions attributed to the prophet Muhammad, which are

revered and received in Islam as a major source of religious law and
moral guidance.

The American Nation of Islam was founded in 1913 in Newark, New
Jersey, by the prophet Noble Drew Ali, born Timothy Drew (1886–1929).

SHINTO

With Buddhism, Shinto is one of the two national religions of Japan. The
name Shinto is derived from the Chinese *shin tao* (way of the gods).
Shinto began around AD 500 and, unlike other religions, has no known
founder, no holy book or scriptures, and no religious law. The priesthood
of Shintoism is loosely organized and preaches nature worship, love of
family, physical cleanliness, hero worship, and shamanism.

The emperor of Japan was always regarded as a god, descended from
Amaterasu the Shinto sun goddess and mythical founder of Japan, until
after the Second World War when Hirohito (1901–89) was forced by the
Americans to renounce his divine status explicitly.

VOODOO

As the national religious folk cult of Haiti, voodoo is followed by at least
80 percent of the population. The word comes from the Haitian *vodun*
meaning a "god" or "spirit." It is a mixture of Roman Catholic
rituals and African magical elements, which were brought to
Haiti by African slaves when it was a French colony. Voodoo
maintains that there is a state between life and death, in
which people become "the undead," which are known as
zombies.

MORMONISM

Mormonism is more properly known as the Church of Jesus Christ of Latter-day Saints.

In 1827, at the age of twenty-two, Joseph Smith (1805–44) began to write the *Book of Mormon* and founded the Mormon faith. The *Book of Mormon* was completed in 1830 and contains more than 250,000 words (a typical novel contains about 70,000 words).

Smith preached widely and sold the *Book of Mormon* in large numbers, so that the Mormon Church acquired many thousands of followers across America. Mormons regarded Joseph Smith as a latter-day prophet at least as important as Moses.

Brigham Young (1801–77), who had joined Joseph Smith in 1833, took up the leadership of the Church, and in 1847 led the people to Salt Lake City, Utah. (At that time, Utah was in Mexico.) Young visited England in 1840–41 to bring the Mormon message to Europe for the first time.

The Church, which is based on Christianity, believes that God's revelations come through modern prophets such as Joseph Smith and Brigham Young, and that the Holy Trinity exists as three separate entities. Mormons refrain from tea, coffee, smoking, drinking alcoholic beverages, and using illegal drugs. Salt Lake City became to Mormonism what Rome is to Roman Catholicism, the de facto capital of the religion.

SCIENTOLOGY

The Church of Scientology began with a best-selling book called *Dianetics: The Modern Science of Mental Health*, written by American science fiction writer L. Ron Hubbard (1911–86), and published in May 1950. Hubbard, the founding father of Scientology, began preaching his first messages in Phoenix, Arizona, in the autumn of 1951, and by the spring of 1952 the new religion had over 15,000 adherents.

Hubbard introduced Scientology as "a study of knowledge," and its founding principles are a form of religious philosophy dedicated, through counselling, to the rehabilitation of the human spirit. Hubbard continued to introduce fresh doctrines for the last thirty-four years of his life. The first church of Scientology was built in Los Angeles in 1954.

From its early days Scientology and its leaders have faced prosecutions for tax evasion, charges of fraud, and accusations of conspiracy to steal government papers. The Church has also often been criticized for the financial demands it puts on its members and what are seen as bogus scientific and religious claims.

MOONIES

More properly known as the Unification Church, the Moonies were founded in Seoul, South Korea, in 1954, by the Reverend Sun Myung Moon, real name Yung Myung Moon (b. 1920).

In 1945 Moon wrote *The Divine Principles*, which became the scripture of the Moonies, but there is evidence to suggest that he stole many of his ideas and teachings from two other cults, which he had earlier been involved with. Moon's overriding claim is that when he was sixteen years old Jesus came to him on a mountain in North Korea and asked him to fulfil His mission on earth. The Unification Church holds many of the traditional Christian beliefs, but also believes that the death of Jesus was not preordained, and that in the Final Days, Satan will become a good angel. The Reverend Moon claims that he is the Messiah of the Second Coming.

Moon achieved success in his quest to build a new religion after the 1961 coup in South Korea, in which the military led by General Park Chung Hee (1917–79) overthrew the civilian government. In 1971, after a couple of periods in jail, Moon moved his base to the United States where the media named his followers Moonies.

The organization owns the *Washington Times* and hundreds of other businesses, including a firearms producer.

ATHEISM

Atheism is the rejection of belief in any spiritual being or the value of any religion.

The majority of atheists base their nonbelief in God or gods on rationalism or philosophy, citing lack of evidence for the existence of a god as the main reason for their lack of belief.

In the pre-Christian era, when the general populace tended to adhere to gods and religions, others of that era refuted the existence of gods. In ancient Greece, the philosopher Socrates (470–399 BC), despite strong evidence to the contrary, was accused of being an atheist and found guilty of insulting Athena, the goddess of wisdom, war, art, industry, justice, and skills. He had actually called into question the ancient Greek conception of gods.

Socrates was sentenced to death by hemlock poisoning, and despite being given the opportunity to escape, accepted his sentence, and insisted on taking the poison.

SEX

COVERING ANCIENT GODS OF SEX, THE *KAMA SUTRA*,
CONTRACEPTION, PROSTITUTION, SEX IN ART,
SEX ON SCREEN, AND HOMOSEXUAL SEX

I know nothing about sex, because I was always married.
—ZSA ZSA GABOR (B. 1919)

ANCIENT GODS OF SEX

Ancient civilizations worshipped gods and goddesses for almost every aspect of everyday life. Sex and love were among the most important.

In Greek mythology Priapus was the god of fertility and male genitalia. Statues of Priapus, showing him with greatly enlarged male genitals, were placed in gardens to encourage crops. (Priapism, named after Priapus, is a painful medical condition in which the erect penis will not return to its flaccid state.)

The ancient Greeks also worshipped Eros, who was their god of lust, love, sex, and fertility, and from whom the word *erotic* is derived.

Aphrodite was the Greek goddess of love, allegedly born in Cyprus. Engaging in intercourse with one of her priestesses was regarded as the same as worshipping Aphrodite.

Roman mythology held that Venus was not only the goddess of love, but also the protector against vice and the goddess of sexual healing. As her natural state was to be naked, Venus became a popular subject for artists.

The Roman equivalent of the Greek Eros was the god of erotic love Cupid, who is mostly portrayed by artists with a bow and a quiver full of arrows, supposedly the arrows of love.

THE *KAMA SUTRA*

Kama is the ancient Indian god of all things pleasurable, including erotic sex, and approximately equivalent to the Roman god Cupid. The word *sutra* means a "short book."

The well-known book, the *Kama Sutra*, which glories in the enjoyment of sex, was written in India some time between the first and sixth centuries by Mallanaga Vatsyayana. The book is mainly about sexual behavior and contains chapters on sexual positions, foreplay, orgasms, seduction, marriage, and many other aspects of sex.

Rather than being merely a how-to manual of all things sexual, the *Kama Sutra* also includes chapters on kissing, courtly behavior, and ways of treating partners, both within marriage and outside of it. It emphasizes the importance of the care and mentoring of partners, encourages mutual experimentation to achieve bliss, and dispenses with Western ideas of guilt, insecurity, or shame in sex.

CONTRACEPTION

One of the earliest contraceptives was used by the ancient Sumerians around 2300 BC. This consisted of balls of opium inserted into the vagina during intercourse and had the added advantage of simultaneously giving the man and woman a "high."

Ancient Egyptian documents record various methods of contraception, including coitus interruptus, although the link between male semen and pregnancy was not fully understood. Around 1850 BC the Egyptians had taken to using crocodile dung, although it is not certain how it was applied.

In Rome in the first and second centuries the Greek gynecologist Soranus of Ephesus (in present-day Turkey) made a study of the subject of contraception and provided a lucid and detailed account of various methods. Hundreds of years ahead of his time, Soranus wrote *Gynecology*, which was the first book to elevate birth control, obstetrics, and gynecology to a legitimate medical speciality.

The Aztecs, who dominated most of the land that is now Mexico from the twelfth century until the Spanish conquest in the early sixteenth century, used a contraceptive made from an unappetizing mixture of eagle excrement and extracts of the pulp of the fruit of the calabash tree.

Condoms have been used for contraception since the seventeenth century. The early versions were made from animal gut or fish membrane.

Giacomo Casanova (1725–98), whose name, along with that of the Marquis de Sade (1740–1814), has become synonymous with sexual excess and depravity, used a squeezed half lemon, placed "strategically," to prevent pregnancies in his many conquests.

Mary Wollstonecraft (1759–97) occupies a very important place in feminist literary studies. Her work in male-dominated eighteenth-century England led to social reforms and the better education of young women, particularly in matters of child rearing and birth control. Wollstonecraft held that "reason" was God's gift to all humans, and in her 1792 book, *Vindication of the Rights of Women*, she made the comparison between the rights of men and the suffering of women, maintaining that women should be equal partners of men in marriage. She also campaigned for women's sexual freedom.

Her 1789 book, *The Female Reader*, a collection of short stories, was published under the pseudonym Mr. Cresswick, "Teacher of Elocution."

Mary Wollstonecraft's daughter was Mary Shelley, the author of *Frankenstein*, which was first published when she was only twenty-one. She was the second wife of the poet Percy Bysshe Shelley.

The first systematic attempt to educate the population of a whole country in methods of birth control took place in Holland in 1882. The program was led by Dr. Aletta Jacobs (1854–1929), the first Dutch woman to attend university and the only female doctor in Holland at that time.

Incurring the wrath of the whole of the Dutch medical establishment, Jacobs began by providing effective contraception to a number of women whose need she deemed to be the greatest. Her work was criticized as threatening to create "a world without children," and she was even criticized from church pulpits by clergymen whose wives were actively consulting her for contraceptive treatment.

Oral contraceptives The three major figures in the development of the oral contraceptive are Margaret Sanger (1879–1966), who in her eighties raised the initial $150,000 to fund the oral contraceptive research project; Frank Colton (1923–2003), who invented Enovid, the first oral contraceptive; and Carl Djerassi (b. 1923), who developed the modern "pill."

Margaret Sanger, an Irish American from a poor working-class family in New York, is widely regarded as the founder of the birth-control movement in the United States. She opened a birth-control clinic in 1916 in New York and published *What Every Girl Should Know*, which provided basic information about topics such as menstruation and sexual feelings. Police were alerted and raided the clinic. Sanger, who had been mailing out birth-control advice, was arrested for distributing obscene material by post, contrary to U.S. Post Office regulations. To escape prosecution, Sanger left America to live in Europe. She returned in 1917, published *What Every Mother Should Know*, and was promptly rearrested and sent to the workhouse for "creating a public nuisance."

> *I was resolved to seek out the root of the evil,*
> *to do something to change the destiny of mothers*
> *whose miseries were as vast as the sky.*
> —MARGARET SANGER (1883-1966)

The pill, as the oral contraceptive is widely known, was introduced to the public in 1961. Women were finally able to take necessary precautions themselves and no longer had to rely on men. The enhanced sensations of condom-free sex, and the combined feelings of emancipation from the straightlaced late 1940s and 1950s, ushered in the freedoms of what is known as the sexual revolution.

The modern IUD (intrauterine device) was pioneered by Richard Richter in Germany in 1909. Otherwise known as "the coil," it was a small metal or plastic device fixed into the uterus. Since 1960 there have been two main types, inert and active. Inert types include the Lippes loop and the Margulies spiral, both of which are made of polyethylene with a barium sulphate addition so that they can be indentified by X-ray. Active IUDs are made of copper with a hormone-releasing agent.

IUDs are usually set into the uterus by means of an inserter tube, which is passed through the cervical canal. The device is secured according- ing to the manufacturer's instructions and may be left in place for up to four years. Removal of the IUD normally al- lows fertility to return.

It is still not certain how IUDs prevent fertility, but it was known in antiquity that a foreign body inserted into the vagina had a con- traceptive effect. Highly polished semiprecious stones were the de- vices of choice for the ancient Egyptians.

PROSTITUTION

*The big difference between sex for money and sex for free
is that sex for money usually costs a lot less.*
—BRENDAN FRANCIS (1923-64)

In Old Testament times Moses (circa 1300 BC) laid down various ordi- nances for the control of venereal diseases since Hebrew law did not for- bid prostitution, it merely confined it to foreign women.

Prostitution is said to be the world's oldest profession and was institu- tionalized in ancient Greek and Roman society. Prostitutes were compelled to wear distinctive dress and pay taxes, unlike in modern Germany where prostitution is officially sanctioned as tax free.

European prostitutes were imported to America during the early 1700s, helping to service the garrisons of soldiers stationed in New York and Boston. It rapidly expanded into other cities from 1810 onward.

Prostitution was legal and virtually uncontrolled in the United States until as late as 1910, when the Woman's Christian Temperance Union began campaigning heavily against it. The Mann Act was passed in 1910, forbidding the transportation of women across state borders for immoral purposes.

The first medical supervision of prostitutes was begun in Buenos Aires in 1875.

SEX IN ART

Ever since writing and painting developed, artists have represented sexual subject matter on walls, paper, and canvas. The ancient Greeks and Romans had few inhibitions when it came to matters of sex, and the depiction of sex in their paintings reflects their relaxed attitudes.

In the four-thousand-year-old Egyptian tomb at Saqqâra, two men, Niankhkhnum and Khnumhotep, are shown sharing a passionate kiss and embrace. In general, ancient Egyptian art was not overtly sexual in content; sex was present more in suggestion than in actuality, especially in high art. The Turin Erotic Papyrus, now housed in the Egyptian Museum in Turin, Italy, is an exception. The papyrus, which is of high artistic merit, was painted in the Ramesside period (1292–1075 BC) and contains a series of twelve vignettes. The vignettes vividly depict a rough-looking man having sexual relations with an attractive, almost naked young woman, in a variety of positions, including standing in a chariot. It is thought that the papyrus was a satire on the prevailing human manners.

The walls inside some of the surviving buildings in Pompeii, which was almost completely destroyed by the eruption of Vesuvius in AD 79, are covered with graphically lewd murals. It is thought that these may have been the walls of brothels.

PORNOGRAPHY

One of the first known writers of sexually explicit literature was Ovid (43 BC–AD 17), who wrote *Ars Amatoria* (*The Art of Love*). In his book, Ovid describes not only the techniques of sex but also advises his readers how often to have sex and how to get the most out of it.

The Italian Pietro Aretino (1492–1556) is regarded as the father of modern pornography. In 1524 he wrote a series of sixteen sonnets, known as "Aretino's Postures," to accompany sixteen erotic drawings of sexual positions, drawn by Giulio Romano, a pupil of Raphael. In the sonnets, he uses highly expressive language to describe the sensations of each position. Aretino was lucky to escape being imprisoned.

SEX ON SCREEN

In 1896, within a year of the world's first public showing of a moving picture, Louise Willy was the first actress to appear naked on screen, in the French film *Le Bain* (*The Bath*). Later the same year, Willy appeared in *Le Coucher de la Marie*, in which she performed a striptease.

Also in the same year, the Catholic Church denounced as pornographic a scene in the filmed version of the stage play *The Widow Jones* showing a couple kissing. The scene lasted for twenty seconds.

The first screen sex goddess was Theda Bara (1885–1955), who appeared for the first time in *A Fool There Was* in 1915.

The first film to show a woman or couple having an orgasm during sex was *Ecstasy*, in 1933, starring Hedy Lamarr (1914–2000). The film was produced in Czechoslovakia, and in 1935 it became the first film to be blocked from entering the United States by U.S. Customs (see "Communication," p. 28 and "War" p. 260).

> *The orgasm has replaced the Cross as the focus*
> *of longing, and the image of fulfilment.*
> —MALCOLM MUGGERIDGE (1903–90)

The Hays Code (also known as the Production Code) was introduced in the United States in 1930, in part to control the depiction of sex on screen. The code laid down rules that restricted nudity, suggestive dancing, ridicule of religion, drug use, and many other aspects of life that could be seen to lower the moral standards of the day. It even included a rule that actors and actresses must both have one foot on the floor at all times during bedroom scenes. The code was strictly enforced after 1934, even prohibiting sexual innuendo, and in doing so severely limited the blossoming career of Mae West (1893–1980). The Hays Code was abolished in 1967.

The first film to be prosecuted under the Obscene Publications Act of Britain was *Last Tango in Paris*, released in 1974, starring Marlon Brando (1924–2004) and Maria Schneider (b. 1952).

Sex and the City, a television program first broadcast on HBO in 1998, is unique in depicting the careers and sex lives of four single women. The women live, work, and play in New York, and the show became aspirational viewing across age groups around the world.

HOMOSEXUAL SEX

The term *homosexual* is generally taken to refer to sex between two males, but strictly speaking it is the overall term referring to sex between any same-sex partners. Sex between females is also referred to as lesbianism. The prefix *homo* is from the Greek meaning "same," not the Latin meaning "man." The first use of the term *homosexual* was in a pamphlet of 1869 written by human rights campaigner Karl-Maria Kertbeny (1824–82).

In ancient Greek mythology, the gods Zephyrus and Hyakinthos were homosexual lovers. They are commemorated in the writings of Homer from the sixth or seventh century BC and on Greek pottery of the same era.

Plato (427–347 BC) and the playwright and politician Sophocles (495–06 BC) wrote extensively in the fourth and fifth centuries BC on homosexual love. Before that, the first great female poet, Sappho, who lived on the Greek island of Lesbos in the seventh century BC, had groups of female admirers and wrote most of her love poems, which were sometimes of a graphic nature, exclusively to women (see "Art," p. 14).

SPACE

COVERING EARLY ASTRONOMY,
TWENTIETH-CENTURY ASTRONOMY, ROCKET SCIENCE,
THE MOON, CONSPIRACIES AND MYTHS,
SATELLITES AND SHUTTLES, SPACE FLIGHT,
DEATHS IN SPACE (AND SPACE PROGRAMS),
AND MISCELLANEOUS

All space is slightly curved.
—ALBERT EINSTEIN (1879-1955)

EARLY ASTRONOMY

SOLAR OBSERVATORY

The first solar observatory was built by Stone Age man at Newgrange, County Meath, Ireland, in 3200 BC. Newgrange, which is reliably dated at six hundred years older than the pyramids, is a vast stone-and-turf mound within a high wall of white quartz. Within the building is a cross-shaped central chamber reached by a sixty-foot-long passageway.

At dawn on the winter solstice, the shortest day of the year (December 22), the first rays of the sun shine along the passage, enter the tomb, and light up the burial chamber for about a quarter of an hour.

PLANETS ORBITING THE SUN, NOT THE EARTH

Accepted church dogma in the fifteenth century maintained that the earth was the center of the universe, and the sun, the planets, and stars orbited around it. Any theory to the contrary was likely to be interpreted as heresy by the Roman Catholic Church, and the perpetrator of any such theory faced being put to death by burning at the stake.

Polish-born Nicolaus Copernicus (1473–1543), who worked in Prussia as a mathematician, economist, and church governor, began his observations of the heavens in 1497. He came to the conclusion that, contrary to Church teaching, in fact the earth and the planets orbited around the sun. He formed the theory of heliocentricity (as opposed to geocentricity), but did not dare publish his work until 1543, the year of his death.

Philolaus (circa 480–405 BC) Although Copernicus was actually the first to publish the theory that the sun is at the center of our solar system, someone else was thinking along the same lines almost 1,800 years earlier. In the fourth century BC, the Greek philosopher and mathematician Philolaus, a contemporary of Socrates, proposed that the earth revolved in a circular orbit around the sun, although he thought the sun was a giant glass disk that reflected the light of the universe. Philolaus was also the first to advance the idea that the earth spins on its axis.

MAPPING THE STARS

The ancient Babylonians are credited with the earliest knowledge of stars and their movements. As early as 3000 BC they recognized and charted the most prominent constellations visible to the naked eye.

The first person to map the stars was Danish astronomer Tycho Brahe (1546–1601), who created a massive observatory on the island of Hven in 1577, nine years before the invention of the telescope.

By observing the motion of a comet in 1577, Tycho (as he is known) determined that another plank of the church's teaching about the heavens was incorrect. It had been the church's position that the moon and all the planets circled the earth, each carried along inside its own "celestial crystal sphere." Because the comet he was observing traveled without hindrance through the area supposedly bound by the so-called spheres, Tycho proved that the spheres did not exist.

The first person to work out the equations for orbiting planets and to show that their orbits were ellipses—flattened circles, rather than perfect circles—was the German Johannes Kepler (1571–1630). Kepler published his findings in *Astronomia Nova* in 1609.

The first person to establish the basic laws of motion, force, and gravity that govern the movement of the planets, and to prove conclusively that Kepler's equations worked, was Isaac Newton (1643–1727). He published his findings in 1687.

TRANSIT OF VENUS

In 1639 the amateur astronomer Jeremiah Horrocks (1617–41) of Toxteth, Liverpool, was the first to observe the transit of Venus, when the planet Venus passes in a direct line between the sun and the earth.

Johannes Kepler had calculated that Venus would slightly miss in its trajectory when crossing the face of the sun. Questioning the calculations, Horrocks made a more accurate forecast of the path Venus would take, and he left a church service to make his observation. He focused his telescope onto a card for thirty minutes, so that the sun could be safely observed. To his relief, his calculation was vindicated when Venus appeared as a tiny black dot moving across the card in front of the sun.

It has been claimed that Horrocks's observation was the most important thirty minutes in the history of man, leading as it did to the accurate measurement of the distance of the sun from the earth, the size of Venus, and other astronomical measurements. Horrocks's measurement of the sun formed the basis for Newton's later work.

TELESCOPES

Galileo Galilei is credited with inventing the telescope in 1609, but the English mathematician and astronomer Thomas Digges is known to have used a device with a convex lens at the front and a reflector at the rear to observe enemy shipping in 1578. He is thought to be the first man to turn a telescope to the skies to observe the stars.

In 1661 James Gregory (1638–75) designed a telescope using mirrors instead of lenses, but did not build it. Five years later Isaac Newton independently developed a fully usable, but small, reflecting telescope.

William Herschel (1738–1822) was born in Hanover, Germany, and emigrated to Britain in 1757 after serving as a bandboy in the Hanoverian Guards. In 1774 he pioneered the use of mirrors in place of lenses in the construction of large "reflecting" telescopes. Reflecting telescopes do not suffer the same optical distortions as refracting telescopes, which use lenses. The lack of distortion enables reflecting telescopes to be constructed on a far bigger scale, thus allowing astronomers to view distant objects much more closely.

Using the new type of telescope, Herschel discovered distant galaxies. He was also the first to observe the clouds of particles now known as nebulae and the first to put forward notions regarding the clustering of nebulae.

Herschel was also one of the first astronomers to develop broad ideas about the nature of the universe. He was the first to propose the theory of stellar evolution—the creation, life, and death of stars—and concluded that the whole of our solar system moves through space.

In 1781 Herschel discovered Uranus, the first new planet to be found since ancient times. As Astronomer Royal, Herschel tried to name the new planet King George's Star after King George III, but it was not accepted. He discovered two satellites of Saturn and two satellites of Uranus. He also discovered infrared radiation and coined the word *asteroid*.

However, even Herschel was fallible. Among his theories was the belief that all of the planets, even the sun, were populated.

FIRST ASTEROID

On January 1, 1801, the first asteroid was discovered by the Italian astronomer and professor of theology Giuseppe Piazzi (1746–1826). Piazzi had stumbled on what we now know as the asteroid belt between Mars and Jupiter.

TWENTIETH-CENTURY ASTRONOMY

EXPANSION OF THE UNIVERSE

In 1920 Edwin Hubble (1889–1953) of the United States was the first to provide evidence of the expansion of the universe. Hubble's constant, which was written in 1929, is a law stating that the speed at which galaxies are drifting apart is constant and has stayed the same for between 10 billion and 20 billion years.

RADIO AND RADAR ASTRONOMY

The birth of radio astronomy occurred in 1932 when twenty-seven-year-old American radio engineer Karl Jansky (1905–50) detected a source of cosmic static. At the time, Jansky was investigating disturbances on the transatlantic telephone cable on behalf of his employer, Bell Telephone Laboratories. He attributed the interference to the interaction between ions and electrons in interstellar space and located the source of the interference as the center of our own galaxy, the Milky Way.

By the mid-1940s astronomers were using large antennae to study faint radio sources from space and to obtain greater detail of the galaxies than was possible by optical observation.

In 1946 astronomers in Hungary and the United States were able to bounce radar waves off the moon and detect the reflected wave. In 1958 radar waves were bounced off Venus for the first time.

Quasars (quasi-stellar radio sources), which are on the very edge of the observable universe and the brightest objects known, were discovered in 1960 by Allan Sandage (b. 1926) and Thomas Matthews using radio astronomy.

Pulsars (pulsating radio stars), which are in fact rapidly spinning collapsed stars, were discovered in 1967 by Jocelyn Bell (b. 1943) while checking massive amounts of printout results from a radio telescope. Since Bell was only a student at the time of her discovery, the 1974 Nobel Prize for Physics was awarded to her Cambridge tutor, Antony Hewish (b. 1924).

Black holes are collapsed stars, which exert such strong gravitational pull that even light cannot escape. Based only on human imagination, the unnamed theory of such entities had already existed for two centuries before the term *black hole* was coined in a 1968 lecture given by American physicist John Wheeler (b. 1906).

The first evidence of a black hole, located in the binary star system Cygnus X-1, was found by NASA's *Uhuru* X-ray satellite in 1972.

The first pictures direct from Mars were beamed back to Earth on July 20, 1976, by the U.S. Mars landing vehicle, *Viking 1*.

The first usable photographs of a comet's nucleus were taken by the European Space Agency's *Giotto* spacecraft as it flew close to Halley's comet in 1986.

Venus was mapped for the first time in September 1994 by the U.S. space probe *Magellan*, using radar imaging techniques. After the completion of its five-year mission *Magellan* was allowed to sink into the dense Venusian atmosphere where it vaporized.

ROCKET SCIENCE

Rockets began as no more than fireworks, but developed into missiles for use in war and ultimately as vehicles to transport men and machinery into space.

The earliest rockets were developed by the ancient Chinese, although the precise date cannot be determined. Gunpowder was packed into an open-ended tube, and as it burned rapidly, the controlled explosion created thrust, causing the forward momentum of the rocket.

The basic equations of rocketry were first calculated by the Russian Konstantin Tsiolkovsky (1857–1935), who completed his work in 1903. He calculated that the escape velocity from the earth's gravitational pull required for a space vehicle to enter orbit was eight kilometers (approximately five miles) per second. He also stated that to achieve the escape velocity, a multi-stage rocket fueled by liquid oxygen and liquid hydrogen would be needed. In his honor, the equation for rocket propulsion is known as the Tsiolkovsky rocket equation. Although regarded as the father of rocketry, and the author of five hundred works on space travel,

Tsiolkovsky spent most of his working life as a high-school mathematics teacher.

The first liquid-powered rocket was produced by the American Robert Goddard (1882–1945) in Massachusetts in 1926. Goddard's first rocket flew just 41 feet high and 184 feet forward, but he had proved his concept would work. For the fuel, Goddard used gasoline and oxygen mixed from separate tanks in the combustion chamber. The immense noise created by his rocket experiments meant that Goddard was forced to move to the town of Roswell, in a remote area in New Mexico, to continue his work.

The U.S. Army failed to grasp the importance of rockets for war purposes.

Leading the German rocket development program was Wernher von Braun (1912–77). Von Braun was captured after the end of the Second World War and offered his services to the United States. He pioneered rocketry for the U.S. space program in the 1950s and 1960s, and saw his Atlas rocket launch a man onto the moon.

The first proton rocket was developed by the Soviet Union in 1965. Proton engines use hydrazine and nitrogen tetroxide, which are hypergolic fuels. Hypergolic fuels burn on contact with each other, saving the need for an ignition system, and reducing the total weight of the rocket.

THE MOON

And there is nothing left remarkable,
beneath the visiting moon.
—WILLIAM SHAKESPEARE (1564–1616)

The first map of the surface of the moon was produced in 1647 by Johannes Hevelius. He was born Jan Heweliusz in Poland in 1611 and became a wealthy brewer and city councillor in Danzig, Germany. He died in Danzig in 1687.

GOING TO THE MOON

Once mankind had mastered the science of rocketry, going to the moon became a real possibility. Stimulated by their competing interests during the cold war, and each with the desire to prove superiority, the United States and the Soviet Union poured massive financial and scientific resources into the pursuit of space travel.

The Soviets were the first to fly a rocket close to the moon, conducting the first unmanned flyby with *Luna 1* on January 2, 1959. Then the Soviet *Luna 2* was deliberately crashed into the moon's surface on September 14, 1959, in the first impact landing. A month later, in October 1959, *Luna 3* took the first picture of the far side—the dark side—of the moon.

It was almost three more years before the United States conducted its first impact landing, on July 26, 1962.

The first scheduled unmanned soft landing on the moon's surface was made by the Soviet *Luna 9* on February 3, 1966. The Soviets had planned to make a soft landing on the moon with *Luna 6* on June 8, 1965, but the spacecraft missed its lunar orbit and flew off into space.

The first U.S. unmanned soft landing on the moon was made by *Surveyor 1* on June 2, 1966. Widespread panic had gripped the U.S. space agency, NASA, which thought it might yet be beaten to landing a man on the moon.

The first humans to fly past the dark side of the moon were Frank Borman, James Lovell, and William Anders, onboard *Apollo 8*, on Christmas Eve 1968. As *Apollo 8* disappeared behind the moon there were some very tense people on the ground in Houston. This was the first time anyone had flown out of radio contact with ground control behind the moon. To put the space capsule into circumlunar orbit, the crew had to fire the engine in a controlled burn while out of radio contact. As the countdown proceeded, tension mounted, but *Apollo 8* emerged into full view at the exact second it had been forecast. Although it did not land on the moon, *Apollo 8* paved the way for all future moon landings.

The first man to step on the moon was Neil Armstrong (b. 1930) on July 15, 1969. Armstrong had a very close brush with death in December

1968. He ejected with only seconds to spare from crashing a lunar landing vehicle, which he was testing.

The first word ever spoken from the moon to earth was "Houston,"
as in "Houston, Tranquility Bay here. The Eagle has landed." Most people
remember Armstrong's words as he stepped from the ladder to the
moon's surface: "That's one small step for [a] man, one giant leap for
mankind."

The first word spoken on the moon (not for transmission to earth)
was "Contact," as in "Contact light," spoken by Edwin "Buzz" Aldrin
(b. 1930). Incidentally, the last word spoken from the moon was "here," as
in "OK, let's get this mother out of here," by Gene Cernan in December
1972.

CONSPIRACIES AND MYTHS

Moonmen In 1835 the *New York Sun* published a lead story claiming
that a new, powerful telescope had managed to focus on small objects on
the surface of the moon. The story claimed that astronomers had seen
strange "man-bat" creatures (half bat, half man) flitting across the land-
scape. In the face of almost universal ridicule, the *New York Sun* was
forced to issue a retraction.

No one has ever been to the moon A 1995 poll by *Time* magazine
revealed that 6 percent of Americans do not believe men ever went to
the moon.

A Fox TV program, broadcast on February 23, 2001, proposed the theory that the whole of the moon-landing program had been filmed in a
studio. This conspiracy theory was based on matters such as the contrast-
ing directions of shadows and the fact that the American flag appeared
to be waving in a breeze. Each claim has largely been discredited, although there has been a surprising lack of response from the North
American Space Agency (NASA).

Urban myth After Neil Armstrong stepped on to the moon's surface,
he is supposed to have said, "Good luck, Mr. Gorski." This refers to a supposed incident during Armstrong's childhood, when he overheard his

next-door neighbor Mrs. Gorski shouting at her husband, "You want oral sex? You'll get oral sex when the kid next door walks on the moon."

Michael Collins (b. 1930) was the pilot of the command module *Columbia* that remained in lunar orbit while Armstrong and Aldrin descended to the moon's surface. On the journey from earth, the crew had debated what Armstrong should say as he stepped from the lunar landing vehicle. Collins is supposed to have said, "If you'd got any balls, Neil, you'd say, 'Oh my God, what is that thing?' Then scream and rip your mike off."

SATELLITES AND SHUTTLES

The first man-made satellite to be launched into space was the Soviet-made *Sputnik 1*, which went into orbit on October 4, 1957. The United States tried to respond by launching *Vanguard*, but it exploded on launch. It was rather unkindly dubbed *Kaputnik*.

The first communication satellite was not, as is commonly supposed, Telstar, but SCORE (signal communication by orbital relay equipment), which was developed by the U.S. Army. SCORE was launched, on board an Atlas rocket, on December 18, 1958, by NASA. The satellite's batteries failed after only twelve days. Telstar was launched in 1962 to relay transmissions between the United States, the United Kingdom, and France.

Arthur C. Clarke (b. 1917), author of *2001: A Space Odyssey*, had predicted the future development of communication satellites in a 1945 magazine article. His idea was that satellites could distribute television programs around the globe. At the time it was considered to be no more than the dream of a science fiction writer, but, true to Clarke's prediction, Telstar was launched in 1962.

The first commercial satellite was *Intelsat 1* (also known as *Early Bird*), which was launched in 1965 by an international consortium led by the United States.

In 1962, NASA planned and designed the first manned, reusable space vehicle, named X20 Dyna Soar, which was to be launched from a Titan 3 rocket. Neil Armstrong was one of the pilots, but the project was abandoned.

The first effective reusable space vehicle, commonly known as a space shuttle, was *Columbia*, which was first launched April 12, 1981. On its twenty-eighth mission in 2003, *Columbia* disintegrated on reentry to the earth's atmosphere and was lost with all hands.

SPACE FLIGHT

The first man in space was the Russian cosmonaut Yury Gagarin (1934–68). On April 12, 1961, he became the first human being to voyage into space on board *Vostok 1*. He completed a single orbit of the earth lasting one hour and forty-eight minutes.

Gagarin died at the age of thirty-four while testing an aircraft. There were strong suspicions that his death was no accident, since he had publicly disagreed with a number of Soviet policies.

The first American in space was Alan Shepard (1923–98) on May 5, 1961, in the *Mercury 3* space capsule, which was launched by a Redstone rocket to an altitude of 116 miles. Shepard's landing point was a mere 302 miles away after completion of this fifteen-minute suborbital flight.

Shepard became the oldest astronaut to walk on the moon, at the age of forty-seven, in February 1971.

The first American to orbit the earth was John Glenn (b. 1921). On February 20, 1962, he orbited the earth several times in *Friendship 7* in a flight of four hours, fifty-five minutes, and twenty-three seconds.

Thirty-six years after his historic first flight, Glenn also became the oldest man into space at the age of seventy-seven. The only ill effect he suffered was that he looked a bit wobbly after landing.

The first woman in space was Valentina Tereshkova (b. 1937). On June 16, 1963, she completed forty-eight orbits of the Earth in seventy-one hours aboard *Vostok 5*. Tereshkova was a textile-factory worker before she enlisted on the Soviet space program in 1962.

The first American woman in space was Dr. Sally Ride (b. 1951). On June 18, 1983, she traveled on board the space shuttle *Challenger*.

The first African American in space was Guion Bluford (b. 1942) on August 30, 1983, on board *Challenger*.

The first Briton in space was Helen Sharman (b. 1963). On May 19, 1991, she became the first British astronaut, on board the Soviet *Soyuz TM-12*.

The first space walk was made by Major Alexei Arkhipovich Leonov (b. 1934) on March 20, 1965. Leonov had been launched on *Voskhod 2*.

The first American space walk was made by Edward White (1930–67) on June 6, 1965. He was later to perish in the *Apollo 1* disaster.

The first space walk without an umbilical was made on February 7, 1984, by Bruce McCandless (b. 1937). Using a manned maneuvering unit (MMU), which was nothing more than a jet pack strapped to his back, he left the safety of *Challenger* on Mission SS-41-B and became the first person to "walk" in space without being tethered.

The first dog in space was Laika, a mongrel stray who had been caught on the streets of Moscow. On November 3, 1957, she was launched into space on board *Sputnik 2*. There was no way to return Laika safely to earth, and she died in space.

The first monkey in space was Gordo, a squirrel monkey. On December 13, 1958, NASA launched Gordo into space aboard a Jupiter rocket. He survived reentry but there was widespread condemnation in the media when he drowned in the ocean after his flotation device failed on landing.

The first animals to return safely from space were Belka and Strelka, two Soviet dogs, who returned safely to earth on August 20, 1960, after a day in space in the company of forty rats.

The first song in space was "Happy Birthday," on *Apollo 9*, in 1968.

The first married couple in space was Mark Lee and Jan Davis in 1992.

The first private spacecraft At 11:08 a.m. on June 23, 2004, *Space Ship One*, designed by Burt Rutan (b. 1943) with pilot Mike Melvill (b. 1941) at the controls, crossed the frontier of space at 62.21 miles above sea level. Melvill was a mere 410 feet above the boundary of the earth's atmosphere, but it was enough to enter the record books.

The first man or woman on Mars. Not yet, not yet! But—he or she is said to be walking on the earth right now. According to NASA's Web site, new space vehicles combining the best of the Apollo and shuttle technology will be going to the moon carrying four astronauts at a time. The moon will be used as a staging post to launch a mission to Mars carrying six crew members. The mission timetable includes returning to the moon in 2018.

DEATHS IN SPACE
(AND SPACE PROGRAMS)

The space programs of the United States and Soviet Union have cost hundreds of lives. It is a feature of these deaths that they have been caused in almost all cases by a set of unique circumstances. Faults have rarely been repeated.

The first rocket death on any space program Officially 92 people, but possibly as many as 150, died on October 24, 1960, in the Nedelin disaster in Russia. This was the first, and remains the largest single space disaster.

The official in charge of the launch, Mitrofan Nedelin, cut corners and ignored safety procedures in a panic to have the launch coincide with the anniversary of the Bolshevik Revolution, and in an attempt to gain favor with Nikita Kruschev (1894–1971), the general secretary of the USSR. There was a major malfunction with the ignition sequence, and the engines fired up as Nedelin and the team of engineers were still on the scaffolding surrounding the rocket. As spectators tried to run they found the tarmac had melted around them. Many became stuck to the ground and unable to move as flames engulfed them.

The first death on a training program On March 23, 1961, Soviet cosmonaut Valentin Bondarenko (1937–61) was killed by a flash fire while training in a ground-based simulator. Bondarenko crater, on the far side of the moon, has been named in his honor.

The first death on the launch pad On January 27, 1967, in *Apollo 1*, Americans Ed White (1930–67), Virgil "Gus" Grissom (1926–67), and Roger Chaffee (1935–67) died in their launch capsule when a fault in the thirty miles of wiring caused sparks, which ignited in the pure pressurized oxygen inside the cabin. The fire spread rapidly in the oxygen-rich atmosphere, igniting the ethylene glycol fuel mixture, and the astronauts were unable to open the escape hatch.

The first parachute malfunction On April 24, 1967, after the successful reentry of *Soyuz 1*, the capsule's parachute failed to open properly at 23,000 feet. The vehicle hit the ground at more than two hundred miles per hour, and the pilot, Vladimir Komarov (1927–67), was killed instantly, becoming the first person to die on a space mission.

The first death on reentry On June 29, 1971, *Soyuz 11* had completed the first successful visit to the world's first space station, *Salyut 1*. The spacecraft was on its final approach to earth, when a valve, just one millimeter in diameter, opened during the reentry phase, allowing air to escape from the spacecraft.

The three crew members were unable to stop the leak, all the air escaped, and the cabin pressure collapsed to zero within two minutes of the valve opening. The craft landed intact, but Georgi Dobrovsky, Victor Patsayev, and Vladislav Volkov had been asphyxiated. They were the first to die at the point of reentry.

The first rocket explosion In 1980 at the Plesetsk Cosmodrome in the Soviet Union, a rocket exploded on the launch pad killing fifty people.

The first death during liftoff was on January 28, 1986, when the U.S. space shuttle *Challenger* exploded seventy-four seconds after liftoff. All seven crew members died.

The first women to die On January 28, 1986, Judith Resnick (1949–86) and Christa McAuliffe (1948–86), crew members on board *Challenger*, died when the spacecraft exploded after launch.

The first private citizen to die on any space program was Christa McAuliffe on board *Challenger* (see above). Christa was a schoolteacher who had applied for the opportunity to be the first private citizen in space and had been selected from 11,000 applicants.

The first American deaths on reentry On February 1, 2003, the U.S. space shuttle *Columbia* broke up on reentry. The *Columbia* Accident Investigation Board found that a piece of insulating foam had broken away on launch and fatally damaged a wing. On reentry, heat entered the wing through the resultant hole and melted the wing from the inside out. The shuttle disintegrated, and all seven crew members died.

The first moonwalker to die was Alan Shepard, the oldest man to walk on the moon at forty-seven and the first American in space. He died of leukemia at age seventy-four on July 21, 1998.

MISCELLANEOUS

ASTEROIDS AND COMETS

The first asteroid rendezvous On February 20, 2001, the U.S. space-craft *NEAR* (near earth asteroid rendezvous) soft-landed on the asteroid Eros after orbiting it for twelve months.

The first impact with a comet was by NASA's deep impact vehicle, 83 million miles away in space. It was deliberately crashed into the Tempel 1 comet on July 4, 2005. The impact velocity was 23,000 miles per hour.

The objective of the mission was to view and analyze the inside of a comet for the first time, and to try to discover the role comets may have played in the formation of the solar system. The experiment has so far revealed the existence of water ice on the surface of the comet and abundant organic material in the interior. It also pinpointed the comet's likely origin, which was the region of space occupied by Uranus and Neptune.

SPACE STATION

The first space station was *Salyut 1*, launched by the Soviet Union on April 19, 1971. It lasted six months, but lost its orbit and disintegrated on reentry into the earth's atmosphere. The few small parts remaining after the disintegration fell into the Pacific Ocean.

SPACE LAW

International Geophysical Year (IGY), a worldwide cooperation between more than two hundred leading international scientists, lasted from July 1957 to December 1958. It coincided with the Soviet Union successfully launching Sputnik and the United States launching Telstar.

Among the objectives of the IGY was to coordinate research into, and observations of, geophysical phenomena such as volcanoes, cosmic rays, earth's magnetism, rocketry, and solar activity. A technical panel was set up to launch the first artificial satellite.

After the two competing launches by America and Russia, the scientists involved in IGY became concerned that the major powers could use space in strategic power games.

The first space law treaty was drawn up in 1967 and signed by sixty-three countries, including the United Kingdom, the United States, and the Soviet Union. Called the Outer Space Treaty, it states that:

- No country is allowed to claim sovereignty over celestial bodies
- No country is to conduct nuclear testing in space
- No country is allowed to launch military action in or from space

SPACE ELEVATOR

In 1895, after visiting the Eiffel Tower in Paris, Konstantin Tsiolkovsky (see above, p. 221) proposed the construction of "celestial castles" in orbit, attached to the earth by vast cables.

The theory of the space elevator, a massively tall tower extending into space, was first proposed in 1970 by Jerome Pearson. While working at the U.S. Air Force Research Laboratory in Ohio, Pearson published his thoughts on a "stairway to heaven" and a "cosmic railway," in the technical journal *Acta Astronomica*. Arthur C. Clarke (b. 1917), author of *2001: A Space Odyssey*, was asked after a talk he had given about the space elevator when he thought construction would be feasible. He answered, "About fifty years after everyone's stopped laughing."

SPORTS

Organized sports formed an important part of the ancient way of life and originated with the Egyptians and Sumerians around 3000 BC. Egyptian games, which are recorded on temple walls, included archery, wrestling, boxing, acrobatics, stick fighting, horse events, and ball games.

The Egyptians used sports for training and strengthening the body as well as for recreation and pleasure. During his reign, the pharaoh Zoser the Great (2667–2648 BC) had himself depicted in a mural taking part in a running competition during the Heb Sed festival in 2650 BC. It was considered important for a pharaoh to establish his supremacy in long-distance races, which could be rough events by today's standards. Permitted tactics included barging and hair pulling.

A mural in a tomb in Saqqâra, which has been dated 2300 BC, shows children taking part in sports.

TRACK AND FIELD

The earliest organized athletic events were held in Egypt and Sumer in southern Mesopotamia in 3000 BC, as described above. The main participants were the ruling elite, and participation was considered essential, particularly in the education of the pharaoh.

Greek games began around 1500 BC, and by the end of the sixth century BC were being held in four main cities: Olympia, the Olympic Games; Delphi, the Pythian Games; Nemea, the Nemean Games; and Corinth, the Isthmian Games.

The first Olympic Games took place in 776 BC. They were considered to be the most important of the four games and were staged every four years for over one thousand years. Because they held such an important role in Greek life, the word *olympiad* was coined to indicate a four-year period. In the early games, participation was restricted to speakers of Greek, all of the contestants were men, and they all contested the events naked. Women were not allowed in as spectators, even though the original Olympic stadium could hold 50,000 people.

We know the name of the first Olympic champion, Coroebus of Elis, who was a cook. Coroebus won the *stadion*, a sprint race of around 207 yards (a *stade*). Hence the track became known as the stadium.

The first modern Olympic Games took place in Athens in 1896 (see "Questionable Origins," p. 186).

FIRST MODERN OLYMPIC ATHLETICS CHAMPIONS
MEN
Track

Event	Winner	Time			Year
		hrs	*mins/secs*		
100 m	T. Burke (U.S.)			12.2	1896
200 m	J. Tewkesbury (U.S.)			22.2	1900
400 m	T. Burke (U.S.)			54.2	1896
800 m	E. Flack (Australia)		2	11.0	1896
1,500 m	E. Flack (Australia)		4	33.2	1896
5,000 m	H. Kolehmainen (Fin.)		14	36.6	1912
10,000 m	H. Kolehmainen (Fin.)		31	20.8	1912
Marathon	Spiridon Louis (Gr.)	2	58	50.0	1896
110 m hurdles	T. Curtis (U.S.)			17.6	1896
400 m hurdles	J. Tewkesbury (U.S.)			57.6	1900

FIRST MODERN OLYMPIC ATHLETICS CHAMPIONS (CONT.)

MEN

Track

Event	Winner	Time		Year
		hrs	*mins/secs*	
3,000 m steeplechase	T. Hodge (GB)	10	00.4	1920
4 x 100 m relay	GB		42.4	1912
4 x 400 m relay	U.S.	3	16.6	1912

Field

Event	Winner	Height/Distance	Year
		Meters	
High jump	E. Clark (U.S.)	1.81	1896
Pole vault	W. Hoyt (U.S.)	3.30	1896
Long jump	E. Clark (U.S.)	6.35	1896
Triple jump	J. Connolly (U.S.)	13.70	1896
Shot put	R. Garrett (U.S.)	11.22	1896
Discus	R. Garrett (U.S.)	29.15	1896
Hammer	J. Flanagan (U.S.)	49.73	1900
Javelin	E. Lemming (Swe.)	54.83	1908
Decathlon	J. Thorpe (U.S.)		1912

The first year women were allowed to participate in official Olympic track and field events was 1928.

WOMEN

Track

Event	Winner	Time		Year
		mins/secs		
100 m	E. Robinson (U.S.)		12.20	1928
200 m	Fanny Blankers-Koen (Neth.)		24.40	1948
400 m	Betty Cuthbert (Australia)		52.00	1960*
800 m	L. Radke-Batschauer (Ger.)	2	16.80	1928*
1,500 m	L. Bragina (USSR)		41.40	1972*
4 x 100 m relay	Can.		48.40	1928
80 m hurdles	M. Didrikson (U.S.)		11.70	1932
4 x 400 m relay	E. Ger.	3	23.00	1972*

Field

Event	Winner	Height/Distance	Year
		Meters	
High jump	E. Catherwood (Can.)	1.59	1928
Long jump	V. Gyarmati (Hun.)	5.69	1948
Shot put	M. Ostermeyer (Fra.)	13.75	1948
Discus	H. Konopacka (Pol.)	39.62	1928
Javelin	M. Didrikson (U.S.)	43.68	1932
Pentathlon	I. Press (USSR)		1964

*Members of the International Olympic Committee were so distressed at the sight of women collapsing on the track in the 800-meter race in 1928 that no further women's events were run at 400, 800, or 1,500 meters until the 1960 Olympics in Rome. The 400-meter and 800-meter events were reintroduced that year, but there was no 1500-meter race until 1972.

The first ever women's modern Olympic champion was Betty Robinson of the United States, who died in 1999.

The International Association of Athletics Federations (IAAF) is the world governing body of athletics and was formed during the Stockholm Olympics in 1912. It was originally called the International Amateur Athletic Federation.

In the United States, the Amateur Athletic Union (AAU) was the governing body of all amateur sports from its formation in 1878 by the runner and rower William B. Curtis until 1979. However, the Amateur Sports Act of 1979 decreed that the AAU could no longer govern more than one sport at the international level, and the Athletics Congress/USA (TAC/USA) took over the role until 1992. In 1992, the name was changed again to USA Track and Field (USATF).

The Amateur Athletic Association (AAA) is the governing body of athletics in Britain and was formed in 1880. The first AAA Championships due to take place that year were rained out.

GOLF

If you watch a game, it's fun. If you play it,
it's recreation. If you work at it, it's golf.
—BOB HOPE (1903–2003)

It is claimed that golf was invented in Scotland, but this is by no means certain—the origins of the sport are shrouded in obscurity. Certainly the oldest golf courses and the oldest golf clubs in the world are in Scotland, and the game is controlled worldwide, except in the United States and Mexico, from St. Andrews in Scotland. The Royal and Ancient Golf Club of St. Andrews, which remains a private members' club, controls the rules of the game, approves changes to the equipment, and runs the Open Championship and Amateur Championship.

Golf has historically been male oriented, hence the myth that the letters of its name stand for "Gentlemen Only Ladies Forbidden."

There are many other contesting claims as to the origin of golf:

Paganica was a game played by the ancient Romans, which involved hitting a wooden or leather ball into a hole with a curved stick. This does sound suspiciously like golf, and *paganica* has perhaps the strongest claim to being the precursor of the modern game. It is said that the Romans brought *paganica* to the countries they conquered throughout Europe in the first century, much as cricket spread throughout the British Empire hundreds of years later.

Kolven or kolf is known to have been played in Holland in 1297, before golf was recorded as a sport. In *kolven*, a curved stick was used to strike a ball into a hole in a frozen lake. There are paintings of people playing the game, and it is mentioned in some documents of the time, but there are no records of the rules.

Chole is another claimant to be the original form of golf. It was played in northern France and Belgium in the thirteenth century, and seems to have existed in different versions. The essence of *chole* was for opposing teams to strike the same ball, but in different directions. Each team could take three strikes at the ball in an effort to score, after which the opposing team could take a single shot to strike the ball in any direction they chose. *Chole* may therefore seem to be more closely related to modern hockey.

Professor Ling Hongling, a Chinese academic based at Lanzhou University, claims that golf was invented in China and was being played in AD 945. The professor bases his claim on a literary reference he has unearthed in a work called the *Dongxuan Records*. The reference is to a pastime called *chuiwan-chui* (hit-ball hit), in which the players hit a ball with a "purposely crafted" stick.

The earliest records of Scottish golf are from 1457, with the proclamation by King James II of Scotland (1430–60) that golf must be banned. The sport, he said, should be "utterley cryed down" because it was interfering with his troops' archery practice.

There are records that golf was being played at St. Andrews in 1552, on the site of what is now the most famous golf course in the world, known as the Old Course.

The first golf club was the Royal and Ancient Golf Club of St. Andrews (known as the R&A), formed in 1754 by "Twenty-two Noblemen and Gentlemen." It was originally known as the Society of St. Andrews Golfers and became "Royal" in 1834, being the second club so honored (Perth was the first). Of the ten oldest clubs, the R&A is the only one that still plays golf over its original links, the Old Course. The Old Course does not belong to the R&A, but the club has certain playing rights granted to it by the owners, the St. Andrews Links Trust.

Only seaside courses are referred to as "links" courses, since they are built on the land that links the sea with the land. By convention, the word *links* is never used for an inland course.

The Royal Burgess Golfing Society of Edinburgh was formed in 1735 (the "Royal" was granted in 1929 by George V), giving it the strongest real claim to be the oldest golf club still in existence. The club plays over the Bruntsfield Links in Edinburgh.

The first golf club in England was the Royal Blackheath Golf Club, which has a tenuous claim to be the world's oldest golf club. The club maintains that it was formed in 1745, and records actually show golf being played regularly on the heath as early as 1608, although this is not claimed as a club event. Royal Blackheath was certainly formed by 1766, making it the oldest club in England.

James I (James VI of Scotland) (1566–1625) brought his clubs to London in 1603 when he was crowned King of England, and he is known to have played golf on the heath.

America's first golf course was built in White Sulphur Springs, West Virginia. Oakhust Links was opened in 1884 by a group of American, Scottish, and English players, and their Oakhurst Challenge Medal is recognized as the first golf prize in the United States. Oakhurst remains in use and is unique in maintaining the traditions of the game by only allowing hickory (wooden) shafted clubs and original "guttie" (gutta-percha) balls. Tees have to be fashioned out of sand, not wood or plastic, and the fairways are kept short by allowing sheep to roam freely.

The first public golf course in the United States was the Van Cortlandt Golf Course, which opened in New York City in 1895. The course remains in operation today, as part of Van Cortlandt Park in the Bronx.

The first major golf championship was the British Open Championship. First played in 1860 at Prestwick, it is the oldest of the major golf championships and is known simply as "the Open."

The first winner of the British Open was Willie Park Sr. (1834–1903) of Musselburgh in East Lothian, Scotland. The championship was played over three rounds of twelve holes each and Park's total score was 174.

The first televised golf tournament was the 1947 U.S. Open Championship, which was broadcast locally in St. Louis, Missouri. The winner was Lew Worsham (1917–90). The first golf tournament broadcast nationally on TV was the 1953 Tam O'Shanter World Championship. The winner was again Lew Worsham, who holed out from 120 yards with a pitching wedge, on the last hole, to win by a single stroke.

RYDER CUP

Arguably the most iconic international match in any sport, the Ryder Cup was first played in 1927, at Worcester Golf Club in Massachusetts, with the United States winning 9½–2½. Samuel Ryder, a wealthy English seed merchant, provided the sponsorship.

Before the Ryder Cup became the official match between the United

States and Great Britain (in 1977 the match was expanded to include Europe), there were a couple of unofficial matches. In 1921 Great Britain ran out the winners by 9–3 at Gleneagles in Scotland, and again in 1926, Great Britain beat the United States by 13½–1½ at Wentworth, England.

During the 1926 match, Samuel Ryder, who was a spectator, sat down in the clubhouse to take tea with the team members of both sides. A proposition was put to him to sponsor a biannual match between the two nations, and to provide a trophy with his name on it. He agreed on the spot, and so, with this gesture, the name of Samuel Ryder passed into history.

The original concept of a match between the professional golfers of the two leading golf countries of their day is credited to James Harnett, a circulation representative with *US Golf Illustrated.* He put forward the idea in 1920, but had difficulties raising the necessary capital until the Professional Golfers' Association of America voted to advance him some funds at their Annual General Meeting in October 1920. Harnett helped to select the team and began the healthy rivalry that continues through to today.

TENNIS

The word *tennis* is derived from the French *tenez* (hold), which was used to warn an opponent that a serve was due. The forerunner of tennis called *jeu de paume* (the game of the palm) was played in twelfth-century France and is still played in parts of Paris. The game involved batting a ball across a net with bare hands, and later, wearing gloves. Rackets were introduced into the game in the sixteenth century. The ball, which was made of leather stuffed with feathers, did not bounce.

Lawn tennis as we know it began in the 1870s: Major Walter C. Wingfield (1833–1912), a British army officer, created what he called sphairistike, as an almost direct crib of the ancient Greek game of the same name. Sphairistike was played on an hourglass-shaped court with a net at head height. In 1865 Wingfield tried to obtain a patent for sphairistike, but failed because the Patent Office recognized that the

French had already been playing the very similar game of tennis for seven hundred years. Far from discouraged, Wingfield spent a good deal of the rest of his life promoting tennis.

The Lawn Tennis Association was formed in 1888.

The first Wimbledon champion was W. Spencer Gore (1850–1906), in 1877. The final was postponed from Monday July 16 to the following Thursday because of rain. Two hundred paying spectators watched the final from a stand made of three planks of wood. Spencer Gore also played cricket for England.

The first woman champion and first French winner was the legendary Suzanne Lenglen (1899–1938), who dominated the championship from 1919 to 1926.

Lenglen fainted and withdrew from the 1926 Wimbledon singles tournament after being informed she had inadvertently kept Queen Mary, who was sitting in the Royal Box, waiting. Lenglen had been misinformed of her starting time.

The first American winner was Bill Tilden (1893–1953), in 1920.

The first African American winner was Althea Gibson (1927–2003), who won the women's championship in 1957 and 1958.

The first Davis Cup match was played between the United States and Great Britain in 1900.

In 1899, four members of the Harvard University Tennis Team came up with the idea of a challenge tennis match between the United States and Great Britain. The trophy was provided by a Harvard team member, Dwight Davis, after whom the trophy was named in 1945 (it was originally known as the International Lawn Tennis Challenge). The first match was held in Boston, Massachusetts, in 1900, with America winning 3–0. In 1905 the competition was expanded, with France, Australasia (combining Australia and New Zealand), Belgium, and Austria invited to send teams.

BOXING

ANCIENT BOXING

There is recent evidence of boxing matches being held in North Africa and Ethiopia as early as 4000 BC, but the most reliable records show that boxing took place in 1500 BC on the island of Crete. There is no written record of Cretan boxing, and most of the evidence is in the form of art. The famous relief on a rhyton from Hagia Triada, dating to the sixteenth century BC, depicts a number of sporting poses including boxing. There is also a fresco at Thera, dated 1550 BC, which shows two young men boxing, with each wearing a glove on the right hand.

The Greeks adopted boxing as an Olympic sport in 688 BC, calling it *pygme* or *pygmahia*.

MODERN BOXING

The first rules for boxing are known as the "London Prize Ring Rules" of 1743, written by Jack Broughton (1703–89). They were used for more than one hundred years.

In 1865 new rules were developed for boxing by John Graham Chambers of England. To add legitimacy to the new rules, Chambers asked John Sholto Douglas, 9th Marquess of Queensberry (1844–1900), to allow his name to be attached to them as sponsor. Queensberry agreed and the new rules were published in 1867. The rules of boxing are referred to worldwide as the Marquess of Queensberry Rules, and Chambers is never heard of.

Until 1892 all boxing contests were conducted using bare knuckles.

The first heavyweight champion, so recognized even though there were no weight divisions at the time, was James Figg (1695–1734) of England. In his whole boxing career Figg only ever lost one bout and claimed to have been ill at the time.

The first world heavyweight champion wearing boxing gloves was James J. Corbett (also known as Gentleman Jim) (1866–1933). He beat

John L. Sullivan (1858–1918), the last bare-knuckle champion, in 1892, after twenty-one rounds, in New Orleans.

The first British-born heavyweight champion of the world was Cornishman Bob Fitzsimmons (1863–1917), who beat James Corbett on March 17, 1897. Fitzsimmons was never more than a middleweight but was successful in the heavyweight division as a result of his ferocious punching strength.

The first African American heavyweight champion of the world was Jack Johnson, who knocked out Tommy Burns in December 1908 in Sydney.

The origin of the expression "up to scratch" comes from early boxing contests, which were conducted outdoors on the ground with no ring. At the conclusion of each round, a scratch would be drawn in the dirt by the referee, roughly in the center of where the contest was taking place. To start the next round, the boxers were obliged to advance to the scratch (come up to scratch) and face each other. If a boxer was unable to make it to the line, he was deemed to be "not up to scratch" and lost the fight.

Subsequently, the term *scratch* was adopted in golf to refer to a skilled golfer who was able to play consistently to par. His handicap was deemed to be "scratch."

HORSE RACING

Racing with horses has taken place since man first managed to ride them and harness them to chariots. Horse racing was included in the ancient Olympic Games in 638 BC.

The origins of horse racing, as we recognize it today, can be traced back to the English knights returning from the Crusades in the twelfth century. None of them came back from the Holy Land empty-handed. Several brought back swift-running Arab stallions and mares, which were crossbred with English horses to add stamina to their speed.

The first recorded horse race in England took place in 1174 at a horse fair in Smithfield, London, although it has been speculated that Roman soldiers raced horses in Yorkshire during the Roman occupation of Britain.

For four centuries, match racing, in which two horses were matched against each other for a wager, was a popular pastime of the wealthy nobility. The sport developed into full professional horse racing between 1702 and 1714, during the reign of Queen Anne (1665–1714). Anne was also instrumental in the foundation of Royal Ascot.

The Jockey Club was formed in 1750 as the controlling body for all horse racing, and remains in control today. In 1793 the Jockey Club authorized the Weatherby family to keep the *General Stud Book* tracing the pedigree of every thoroughbred horse in Britain. The family has continued the tradition to this day, and the origin of every thoroughbred horse running in Britain can be traced back to just three stallions, called the Foundation Sires: the Byerley Turk of 1679, the Darley Arabian of 1700, and the Godolphin Arabian of 1724.

The Derby was first run in 1780. It was named after Edward Smith-Stanley, 12th Earl of Derby (1752–1834). At a celebration after the Epsom Oaks in 1779, the Earl of Derby and Sir Charles Bunbury tossed a coin to see who would have the race named after them. Derby won the toss, but Bunbury won the first race in 1780, collecting a prize of one thousand pounds. Derby won his own race in 1787.

The first organized sporting event in the United States was horse racing, on Long Island in 1664.

SOCCER

It has now been officially recognized by the governing bodies that soccer began in China more than 2,500 years ago. The ancient Chinese played a form of soccer known as *cuju*, which was popular between 770 and 476 BC.

It originated in the Linzi district of the city of Zibo, on the Shandong Peninsula. The rules involved kicking an inflated pig's bladder around an enclosed courtyard, and the ball had to be kept in the air—a score could not be made if the ball had gone to ground. There were several versions of the game, and one even had a goal net similar to the modern one.

The first set of rules that would be recognized today was drawn up in 1848 at Trinity College, Cambridge. Shin kicking, known as hacking, was allowed as a legitimate means of tackling an opponent. The word *soccer* is an abbreviation of *association*, as in association football. The sport is known as football outside the United States.

The first professional football club, Notts County, was formed in 1862.

The first soccer organization in the United States was the Oneida Football Club. It was formed in Boston in 1862 by seventeen-year-old Gerritt Smith-Miller, and the first game was played on November 7, 1863. In the 1863–65 seasons Oneida did not lose a game and did not concede a single goal.

Referees were introduced to football in 1871 to avoid what had become regular disputes in professional matches.

Fédération Internationale de Football Association (FIFA), the international ruling body, was formed in 1904.

The first FA Cup final at Wembley Stadium was played in 1923. Estimates vary, but as many as 200,000 spectators may have been crammed in.

The first World Cup between nations was won by Uruguay, which beat Argentina 4–2 in the final in 1930. Neither side could agree on the size of ball to be used in the final. A compromise was reached by using one size in the first half and another in the second.

RUGBY

The myth endures that rugby was born as a result of a single incident in 1823. During a game of the very individual version of football being played at Rugby School, William Webb Ellis (1806–72), a pupil, supposedly picked up the ball and ran to the goal line. No hard historical evidence exists to support this romantic story. Nonetheless, the Rugby Union World Cup is named the Webb Ellis Trophy in commemoration of the event and the man.

The rules of rugby were first codified in 1845 by three pupils of Rugby School. Until that time, opposing teams would meet to agree on their own set of rules before taking to the field. Inevitably, lack of a fixed set of rules led to disputes and occasional brawling.

In the beginning, a try was called a "run-in," and no points were scored. Touching the ball down over the try line allowed the team to "try" to convert the touchdown into a goal.

In 1884 the scoring system was changed so that an unconverted try was valued at one point, which could be converted into three points if the goal was scored.

American football, Canadian football, and Australian rules football are all descended from rugby.

SWIMMING

The earliest evidence of man's ability to swim is shown on cave drawings, which have been found in the so-called cave of swimmers at Wadi Sora in Egypt, dated at up to 14,000 years old. An ancient Egyptian clay seal, dated between 9000 and 4000 BC, shows four swimmers using the stroke now known as the front crawl or freestyle.

The earliest written references to swimming date from 2000 BC, occurring in *Gilgamesh*, the *Iliad*, the Bible (Ezekiel 47:5), and the *Beowulf* sagas.

The first book on swimming was *Colymbetes*, written in 1538 by Nicholas Wynman, a German language professor, mainly as a means of reducing the dangers of drowning.

The first swimming competitions were held in England before 1837. At that time there were six man-made swimming pools in London.

The front crawl was originally known as the Trudgen on account of being developed by Englishman John Trudgen in 1873. Trudgen had copied the style from Native Americans after seeing its use on a trip to

South America, although it had been seen in 1844 in London, also being used by Native Americans. This style was thought of as un-English, as the leg kicking created far more splashing than the then current breaststroke.

MILESTONES

The first person to swim the English Channel was the Englishman Captain Matthew Webb (1848–83), in 1875, taking twenty-one hours and fifteen minutes to swim from Dover to Calais. Captain Webb died after being sucked into a whirlpool, in the rapids at the foot of Niagara Falls. He was trying to win a prize of £12,000 in what was considered a suicidal feat.

The English Channel was not swum again until 1911, by Englishman T. W. Burgess.

The first woman to swim the English Channel was Olympic gold-medalist Gertrude Ederle (1906–2003) of the United States, in 1926.

OLYMPIC SWIMMING

Although the Greeks did not include swimming in the ancient Olympics, it was included at Athens in 1896 in the first of the modern Olympics. Only four swimming events were contested in 1896: the 100 meter, 500 meter, and 1,200 meter (all freestyle), and the 100 meter for serving sailors!

The first Olympic swimming gold medal was won by Alfréd Hajós (born Arnold Guttmann) (1878–1955) of Hungary in the 100-meter freestyle in 1896. Hajós also played soccer for Hungary and won the Hungarian 400-meter hurdle and discus championships.

CYCLING

The controlling body of world cycling is the Union Cycliste Internationale, based in Switzerland, which was founded in 1900.

Velocipede racing began in France in 1867. The velocipede was the forerunner of the bicycle.

The Tour de France was first staged in 1903, allegedly by Henri Desgrange (1865–1940) as a publicity stunt to boost the circulation of *L'Auto*, a national newspaper. The idea came from a journalist on *L'Auto*, Géo Lefèvre. The race was won that year by Maurice Garin (1871–1957) of France, with a winning margin of two hours and forty-nine minutes, which remains the greatest winning margin in the history of the Tour de France.

In 1994, in an unparalleled act of entente cordiale, to mark the opening of the Channel Tunnel, part of the Tour de France took place in England.

MOTOR RACING

In 1887 in Paris, the newspaper *Le Velocipede* announced plans for the first motor car "reliability trial." Only one competitor showed up and the race was canceled. A trial finally took place in 1894, with cars racing between Paris and Rouen. This time twenty-one competitors took part, and the trial was won by Le Comte Albert de Dion on a steam-driven tractor. The tractor traveled at an average speed of eleven-and-a-half miles per hour, including a break for a spot of lunch.

The first motor race in the United States was won by Frank Duryea (1870–1967), over a fifty-four-mile course in Chicago in 1895.

The first Grand Prix race was the French Grand Prix. The race was held in 1906 over 686 miles on a 60-mile road circuit, close to the town of Le Mans. The first winner was the Hungarian driver Ferenc Szisz (1873–1944), driving a Renault AK 90CV.

The world's first off-road race track built for that purpose was Brooklands at Weybridge in Surrey, which was completed in 1907.

The Monte Carlo Rally first took place in 1911. It was won by Henri Rougier in a Turçat-Méry.

BASKETBALL

Uniquely among major sports, basketball has not evolved and developed out of a group activity with long historical roots, but was thought up by a single person, acting alone.

In the winter of 1891 Dr. James Naismith (1861–1939), having been appointed athletic director at the YMCA Training School in Springfield, Massachusetts, was faced with the problem of finding a sport to play indoors during the cold winter months of that northern town. He wanted the sport to require skill, mobility, and ingenuity, rather than merely to be a test of strength.

Naismith set out to create a game that could be played on a small area, and within fourteen days had managed to devise a set of rules governing what would become basketball.

The first basketball games took place in late 1891 and were played with a soccer ball. The baskets, which were provided by the school janitor, were peach baskets with closed ends. This meant that play had to stop for the ball to be recovered from the basket, by hand, after every goal.

Basketball's popularity spread rapidly through the YMCA movement in many nations, and was introduced into the Olympic program of 1936.

The National Basketball Association (NBA) was formed in 1946, originally as the Basketball Association of America (BAA). The name was changed in 1949 after teams from the National Basketball League (NBL) joined it.

BASEBALL

In 1744 the English author and book publisher John Newbery (1713–67) produced *A Pretty Little Pocket-Book*, in which he referred to a game he called "base-ball." As in modern baseball, a pitcher had to throw a ball to a batter, who tried to hit it. To score a "run" the batter had to run to a base and return. The book was reprinted in the United States in 1762.

The traditional story of the origin of baseball is that in 1839 Civil War General Abner Doubleday (1819–93) formulated the first set of rules for what he called "town ball," in Cooperstown, New York. Although this story is generally accepted within baseball, it is unlikely to be true. Baseball is thought more likely to have developed out of the English game of rounders, which itself developed out of earlier English games such as "stool ball."

Doubleday's nickname in the U.S. Army was "forty-eight hours" (a double day). He is noted for firing the first cannon shot in defense of Fort Sumter in the Civil War. After the Civil War had ended, Doubleday bought the company that still operates the cable cars in San Francisco.

The American Civil War of 1861–65 witnessed the rapid growth in popularity of baseball among the serving troops, and within ten years of the end of hostilities, professional players began to emerge. The National Association of Professional Base Ball Players was formed in 1871.

The original five members of the Hall of Fame of Baseball, elected in 1935, are Walter Johnson, Christy Matthewson, Babe Ruth, Honus Wagner, and Ty Cobb.

DARTS

Played mainly in the public bars of English public houses, darts is another sport with an uncertain origin. One theory is that soldiers passed time by throwing shortened arrows at the bottom of upturned wine casks. This speculation has some merit, as it is known that English archers used a form of darts as a training aid during the Middle Ages, and that later Henry VIII (1491–1547) played darts. The numbering of the board in its current layout was devised in 1896 by Brian Gamlin.

The National Darts Association was founded in 1953.

ARCHERY

Recreational archery was practiced by the ancient Egyptians, although there is no record of any competitions. The ancient Greeks also used archery as a form of recreation.

The earliest book on archery was *Toxophilus, or the Schole or Partitions of Shooting*, which was written by Roger Ascham and published in 1545.

The first recorded archery competition took place in 1583 in England, and archery has been an Olympic event since 1900. The modern sport of archery developed out of the military practice of shooting at targets to sharpen the archers' skills.

BILLIARDS

The earliest European reference to billiards is in the fifteenth century, but very little is certain about the origins of the sport. Its beginning has been credited to China, France, Italy, Spain, and England. All the games are played on a table with six pockets except French billiards, which is known as Carom and which has no pockets on the table. The first professional English billiards tournament was won by Jacob Schaefer Sr. in 1879. He managed to work the balls into a corner and amassed 690 points by making repeated contact with the balls while hardly moving the cue ball. This type of play was later outlawed.

The most popular variation of billiards now played in America is eight-ball pool.

WAR

COVERING POISON GAS AND STINK BOMBS,
ANCIENT PROJECTILES AND LAUNCHERS, ROCKETS,
GUNPOWDER AND GUNS, WAR AT SEA, TANKS, RADAR,
AIR COMBAT, NUCLEAR WEAPONS, WAR CRIMES,
AWARDS, ORIGINS OF SOME WARS, ESPIONAGE,
TERRORISM, AND MISCELLANEOUS

Mankind must put an end to war,
or war will put an end to mankind.
—JOHN F. KENNEDY (1917-63)

POISON GAS AND STINK BOMBS

The Hague Convention of 1899 banned the use of specially designed shells for the dissemination of poison gas. Even earlier, the Brussels Declaration forbade the use of poison weapons, although it is not clear if this refers to gas.

POISON GAS
It is a common misconception that the German mustard gas attacks of the First World War signaled the first use of lethal gas as a weapon of war. More than four hundred years earlier, Leonardo da Vinci (1452–1519) had devised a form of shell for Ludovico Sforza, the Duke of Milan

(1452–1508), to use in defense of his city. The shell, containing powdered arsenic and powdered sulphur, would explode on landing to create a poison gas cloud. It is not known if the shells were ever deployed.

Even earlier, the first recorded use of poison gas was in the Peloponnesian War, 2,400 years ago, when the Spartans used arsenic smoke during the sieges at Plataea in 429 BC and Delium in 424 BC.

The first full-scale use of poison gas in a modern war was the German use of chlorine gas on April 15, 1915, during the second battle of Ypres. The French had used tear gas earlier in the conflict, but in a limited deployment.

STINK BOMBS

In 80 BC the Romans used toxic smoke, a filthy combination of urine, rotten eggs, and beer, in a battle against the Charakitanes in Spain. As well as smelling foul, the smoke caused pulmonary problems and blindness, leading to the defeat of the Charakitanes within two days.

Only in the twentieth century were stink bombs used as a form of practical joke.

ANCIENT PROJECTILES AND LAUNCHERS

BOW AND ARROW

Prehistoric bows and arrows have been found in every part of the inhabited world with the exception of Australia.

The earliest evidence of bows is in Tunisia, where a 50,000-year-old bow has been discovered.

The English-Welsh longbow played a key role in English military dominance in the Middle Ages. It was used extensively in the Battle of Agincourt, which famously took place on Saint Crispin's Day 1415, between the English led by Henry V (1387–1422), and the French led by Charles VI (1368–1422) (also known as Charles the Mad). The victory at

Agincourt, which was part of the Hundred Years' War, enabled Henry to claim the throne of France.

The V sign used as an insult originated on the battlefield of Agincourt. English bowmen gave the sign to their French opponents to show that their index and middle fingers were intact, after the French had started the punishment of cutting those fingers from captured enemy bowmen.

The first use of the longbow in a significant battle was at the Battle of Crécy on August 26, 1346. Using bodkin arrows, which had squared metal spikes for arrow tips, the English archers were able to pierce the body armor of the advancing French knights, and as many as 1,200 were killed in this way.

CROSSBOW

The earliest record of crossbows used in war is at the Battle of Ma-Ling in China in 341 BC.

Catapulta

The ancient Romans used the *catapulta* to fire arrows and darts at enemy armies. It resembled a large horizontal bow, about six feet from tip to tip and standing shoulder height to a centurion. The arrows varied from approximately eighteen inches to more than three feet in length.

Even earlier, from 375 BC, the ancient Greeks had a form of *catapulta* called the *oxybeles*, which translates literally as the "bolt shooter." In 400 BC they had the *gastrophetes* (belly bow), which was braced against the abdomen.

Ballista

The term *ballista* includes all forms of catapult-type weapons and was used by Philip II of Macedon. The Romans also used the *ballista* as one of their principal siege weapons. It was a larger machine that hurled heavy stones into enemy strongholds. Both the *catapulta* and *ballista* used the power of the sudden release of tension on the throwing arm. Tension in the *ballista* was created with the use of cords made from wound horsehair, animal gut, or sinews.

Trebuchet

It is thought that a machine similar to the trebuchet was first developed in China around 400 BC and by around AD 500 had been introduced into Europe.

The medieval variant of the *ballista* and *catapulta* was the trebuchet, which acted somewhat like a huge slingshot. It used heavy counterweights on the throwing arm to create the throwing power. The missile was carried in a net suspended with rope from the end of the throwing arm. The additional slingshot effect created extra throwing power. The trebuchet was mainly used as a siege weapon to break through enemy walls and employed a large rock or metal ball for ammunition.

ROCKETS

An early form of rocket was seen in a toy developed by Archytas in ancient Greece in about 400 BC. Archytas suspended a wooden pigeon on a wire and arranged for escaping steam from its tail to propel the bird around in a circle, to the utter amazement of his audiences.

The first use of rockets in a theater of war was at the battle of Kai-Keng in AD 1232, between the Chinese and Mongol invaders. The Chinese called their rockets "fire arrows" and launched them in an attempt to have a psychological effect against the opposition troops. The rockets were simple paper and shellac tubes filled with gunpowder, open at one end. When ignited, the explosive effect of the gunpowder's rapid burning caused the rocket to be propelled forward with great speed. The Mongols eventually prevailed over the Chinese, but were impressed with the effect of fire being delivered by air and rapidly began to develop their own rockets.

The first rocket attack in Europe was made by the Mongols in 1241 at the battle of Legnica in Poland, the furthest west the Mongols fought. Very little is known of the course of the battle except that the Polish leader, Henryk II, Duke of Silesia was killed, and that the Mongols withdrew.

V-1 (FZG 76) rockets were test launched in 1942 and first launched offensively by the Germans on June 12, 1944, against England. Although the V-1 is refrerred to as a rocket, it was propelled by an air-breathing pulse jet.

V-2 The *Vergeltungswaffe 2* (reprisal weapon 2) was the world's first ballistic rocket. A ballistic missile is one that follows a prescribed course that cannot be significantly altered once the fuel is spent, and whose flight is governed by the laws of ballistics.

Testing of the V-2 began at Peenemunde, the German rocket test center, in March 1942 with a spectacular explosion on the launch pad. Testing continued with equally disastrous results until the first partial success in October 1942. Mass production of the V-2 began in 1943 in an underground works near Nordhausen in Germany.

The first successful V-2 offensive launch was on September 8, 1944, against Paris. The final tally of V-2 rockets launched against England was more than 1,400.

The V-2 rocket has a unique characteristic in the history of arms. More deaths were caused by its production than by its military use.

GUNPOWDER AND GUNS

GUNPOWDER

It is thought that gunpowder was invented in China in the eleventh century. The English scientist and Franciscan friar Roger Bacon (1214–94) is also credited with its invention; the formula was discovered in his papers after his death.

The first large-scale facility for gunpowder production was set up in England in 1865 by the Grueber family, who were Huguenot refugees from France. Most of the production was sold to the monarch.

In a tragic twist of fate, Grueber's son was killed as a result of an explosion in the factory. He was out boating on a nearby lake when the explosion happened. Falling debris landed on the boat and killed him.

Human urine was a vital ingredient in the best gunpowder, and in 1626 a statute was imposed by King Charles I that compelled the storage and collection of urine from households every three months. Failure to supply urine put one at risk of severe punishment.

It was wrongly thought that only male human urine would be suitable as an ingredient for the best gunpowder. Nevertheless, during the American Civil War, the Niter and Mining Bureau appealed to "The Ladies of Selma to preserve all their chamber ley collected about their premises for the purpose of making Niter." Niter was another name for saltpeter (potassium nitrate), a component of gunpowder. "Ley" was urine. This request inspired a number of waggish rhymes such as:

> We think the girls do work enough, in making love and kissing,
> But now you've put the pretty dears, to patriotic pissing.

GUNS
Handguns
The first handheld gun was the cumbersome, slow-firing matchlock, which appeared in 1450. To fire the ammunition, a length of match cord was used to ignite powder in a flash pan (the expression "flash in the pan" came from this), which in turn ignited powder in the barrel. This then propelled a lead ball in the general direction of the target.

The system had a significant number of flaws, not least of which was the need to keep an open spark close to the powder. This allowed the enemy to pinpoint the soldier at night, and the high number of accidental discharges proved more dangerous than facing the enemy.

The wheel lock of 1517 was an improvement on the matchlock, and then came the wonderfully named snaphaunce of 1570, an early version of the flintlock.

The flintlock was introduced in 1617. It incorporated the first reliable mechanism for firing a gun, and became the weapon of choice for two hundred years, until self-contained, or cartridge, ammunition was developed. The main design advance of the flintlock involved the use of flint, an extremely hard rock, to strike a spark. A small piece of flint was held in a spring-loaded striker (also called the cock, hence "cocking a gun"). Pulling the trigger released the cock, which struck the flint against a steel striking plate, known as the "frizzen." The resultant spark ignited gunpowder loaded in the barrel, which propelled a lead ball, or, sometimes, round shot. The flintlock had an effective killing distance up to one hundred yards.

The revolver was developed by Connecticut-born Samuel Colt (1814-62). Legend has it that Colt, who was working as a deckhand on a ship, observed the way the capstan worked to lift the anchor and formed the idea of a self-loading firearm.

The unique feature of Colt's design was that six rounds of ammunition were held ready in a cylindrical revolving magazine, each round to be fired through a single barrel. Previous attempts at automatic loading had involved the use of revolving barrels and a single loading mechanism. The Colt design saved both cost and weight, and was far more reliable. Colt patented his idea in England and France in 1835, and in the United States in 1836.

The production line and interchangeable parts were developed by Colt with the help of Eli Whitney (1765-1825) in order to produce the vast number of revolvers needed by the U.S. Army in the 1845-48 war with Mexico.

The first silencer for revolvers was invented by Hiram Maxim (1840-1916). He called it the "suppressor" and was granted a patent in 1909.

The first gun specifically designed to be fired from the shoulder was the Spanish harquebus (also known as the arquebus or hackbut), which was designed in 1450. This forerunner of the modern rifle was effective up to 660 feet.

Cannon

Crude cannons were first used on a European battlefield in 1327 by Edward III of England (1312-77) in his military actions against the Scots.

Machine guns

James Puckle (1667-1724), a London lawyer, patented a tripod-mounted machine gun with a revolving ammunition cylinder in 1718. It fed rounds of ammunition into the gun's single chamber and was capable of firing nine shots without reloading. The weapon failed as a result of its unreliable flintlock firing system. Puckle tried and failed to raise funds to mass-produce his guns. One newspaper neatly observed, "Those are only wounded who hold shares therein."

The first successful machine gun was the hand-cranked Gatling gun, which was developed during the American Civil War by Dr. Richard Gatling (1818–1903) and first used in 1861. The Gatling gun used cartridges that contained the primer, propellant, and bullet. It could fire two hundred rounds per minute.

The first truly automatic machine gun was developed by Hiram Maxim (1840–1916) in 1884. Maxim patented his machine gun, which harnessed the recoil force from the firing of each bullet to work the bolt, expel the spent cartridge, and load the next bullet. The Maxim gun could fire five hundred rounds per minute. Maxim was born in America and emigrated to England in 1881. He became a naturalized British citizen in 1900 and was knighted by Queen Victoria in 1901.

Grenades

Invented in the fifteenth century in France, grenades were so named after the early models' resemblance to pomegranates. The French word for pomegranate is *grenade*.

Soldiers who were specially trained to throw grenades were called grenadiers.

The grenade went out of fashion and was almost unused during the nineteenth century, but its use was revived during the Russo-Japanese War of 1902 and the First World War.

The grenades used during the early months of the First World War were as dangerous to the soldier throwing them as to the enemy, because the handler could catch the rear of the trench during the throw causing the grenade to fall back and explode in the trench.

The first safe grenade was the British Mills bomb, named after William Mills (1856–1932) of Birmingham, who developed it. The Mills bomb was first used on the front line in the First World War in May 1915. It looked more like a pineapple than a pomegranate. Over 70 million Mills bombs were issued to soldiers up till 1970.

SHRAPNEL

Sir Henry Shrapnel (1761–1842) was appointed inspector of artillery in 1804 and promptly invented the shrapnel shell. The artillery shell case contained fragments of metal and was designed so that upon striking its target it exploded, scattering lethal red-hot metal over a wide area. The idea was to wound and disable, as much as to kill.

WAR AT SEA

SUBMARINES, TORPEDOES, AND DEPTH CHARGES

Far from being a modern concept, undersea warfare originated centuries ago when the dominance of the surface-treading warship was challenged from below the waves.

Submarines

The principles of the submarine were first accurately described in 1578 by William Bourne (1535–82), an English mathematician and innkeeper. His design was for an underwater rowing boat, covered in waterproof leather, but not one was ever built.

The first submarine to be built and hold successful trials was by Cornelius Drebbel (1572–1633). He was born in Holland but lived in England from 1604. Between 1620 and 1624 he built the first submarine and held successful trials in the Thames. Drebbel's vessel was propelled by twelve oarsmen, and there is a possibility that James I (VI of Scotland), who was Drebbel's patron, took a short ride in it, becoming the world's first monarch to travel underwater.

The invention of goatskin ballast tanks in 1747 overcame the problem of controlling descent to significant depths.

The first submarine to be used in war was the *Turtle*. In 1776 David Bushnell (1742–1824), a student at Yale University, designed the *Turtle*, which was built by the U.S. Navy to his specification, using a screw propeller. Sergeant Ezra Lee sailed it close to a British battleship in New York harbor, planning to attach explosives to the hull by drilling into it. The hull of the British ship was too tough and the attempt failed, but this was the first time a submarine had been used in war.

The first submarine to deliver and detonate a torpedo successfully was the Confederate CSS *H. L. Hunley*, in 1864, during the American

Civil War. The torpedo, which had been attached to the bow of the *Hunley*, sank its target, the USS *Housatonic*. The *Hunley* also sank in the same action, with the loss of all on board.

In 2000 the *Hunley* was raised from the seabed and is now in the process of restoration.

Torpedoes

The torpedo is named after the torpedo fish, which disables its victims with an electrical discharge.

The modern self-propelled torpedo was developed by Robert Whitehead (1823–1905), a British designer working for the Austrian Navy. By 1866 he had produced a successful working prototype, with an explosive charge in the nose.

The modern torpedo guidance system was invented in 1942 by Hollywood beauty Hedy Lamarr (1914–2000). She patented the invention, which was based on frequency switching, but the patent ran out before it was taken up by the military.

The same system is used today in mobile phone technology (see "Communication," p. 28 and "Sex," p. 214).

Depth charges

The main weapon used by surface vessels against submarines are depth charges. They are in effect waterproof bombs dropped into the sea and set to explode at a predetermined depth. Depth charges were developed by the British navy in its war against the German U-boats during the First World War, and were first used in 1915.

TANKS

The modern tank was developed in 1914 by Sir Ernest Swinton, DSO (1868–1951), who was an official British war correspondent on the western front during the First World War. Swinton was alarmed at the high casualty rate of frontline soldiers, who were being cut down by the thousands by machine-gun fire. Taking his inspiration from a Holt's tractor, which was a heavy caterpillar-tracked vehicle, Swinton wrote a strong memo to the secretary of the War Council in London, urging him to develop a protected means of transport.

The British Landship Committee was set up by the government and agreed to adapt tractors for military use. The resultant vehicles, known as tanks, were used for the first time on August 15, 1916, in the first battle of the Somme.

Leonardo da Vinci, whose work preempted the designs of many later inventions, designed a self-propelled tank nearly five hundred years earlier.

RADAR

The use of radar (radio direction and ranging) was first developed by the German physicist Heinrich Hertz (1857–94) in 1887. Hertz conducted a series of experiments with radio in his laboratory and found that, while the radio waves would pass through certain materials, other materials reflected the waves, creating a sort of echo.

Radar's practical use for the detection of ships and aircraft that were otherwise invisible to the naked eye was refined during the Second World War. Late in the war, British fighter aircraft were equipped with portable radar sets, which enabled them to locate the enemy well before they were seen themselves. It was fed back to the British that German high command was becoming increasingly frustrated that its aircraft were being shot down in ever larger numbers at night.

Seizing the opportunity to spread disinformation, British counterintelligence leaked a story that British pilots were eating a diet high in carrots to improve their eyesight in the dark. This story put the Germans off the scent of mobile radar for nearly a year (see "Inventions," p. 153).

AIR COMBAT

Airplanes are interesting toys, but of no military value.
—FERDINAND FOCH (1851–1929)

The world's first fighter aircraft was the Vickers EFB1 (short for experimental fighting biplane), nicknamed the Destroyer, which first flew in 1912. The EFB1 was exhibited in 1913 at the Olympia Air Show and went into service with the Royal Flying Corps in 1914 as the Vickers Gunbus. The Destroyer was armed with a Maxim .303 caliber machine gun on a swivel mount, and saw its first action on Christmas Day 1914 when it intercepted and shot down a German monoplane.

The first aerial dogfight took place on October 5, 1914, when a French Voisin 3, armed with a Hotchkiss 8 mm machine gun, shot down a German Aviatik B1.

The world's first bomber aircraft was the German airship zeppelin, named after its pioneer Count Ferdinand von Zeppelin (1838–1917). The zeppelin began its role as a bomber in August 1914, when it bombed military targets in Liège in Belgium.

The first time civilians were bombed was on January 19, 1915, when a zeppelin bombed Great Yarmouth. Four people were killed and sixteen injured. It was also the first time the British mainland had been bombed from the air.

London was bombed for the first time on May 31, 1915, also by a zeppelin.

The first time a bomb was dropped by aircraft was in 1911 when an Italian pilot dropped four hand grenades on Turkish targets in Libya. However, the aircraft was on reconnaissance and not strictly recognized as a bomber.

The first aircraft launched from a ship was on November 4, 1910. To test the effectiveness of ship-launched aircraft, Eugene Ely (1886–1911), a civilian pilot who had learned to fly only earlier that year, was launched off a platform on the USS *Birmingham*.

The first successful landing of an aircraft on a ship was also by Ely when he managed to land safely on the quarterdeck of the battleship, USS *Pennsylvania*, on January 18, 1911. The arresting gear on the *Pennsylvania*, designed to stop the aircraft from shooting off the far end of the deck, was constructed of wire attached to heavy sandbags.

SPITFIRE

The Supermarine Spitfire became the icon of British resistance to the Nazi airborne invasion during the Second World War. It was designed by Reginald Mitchell (1895–1937) in response to the growth of Germany's Luftwaffe, and the first prototype, the F37/34, flew in March 1936. The Spitfire entered Royal Air Force service in 1938, and by the outbreak of war in 1939, the RAF had 2,160 on order.

The first vertical takeoff and landing (VTOL) aircraft with fixed wings, rather than a rotor, was the AV8A Harrier jump jet. It first flew on August 31, 1966, and entered service with the RAF on April 1, 1969.

STEALTH

It has been found that certain shapes show up more easily on radar screens. Stealth technology employs radical changes to the designs of both airframes and engines that enable aircraft to fly missions into highly defended target areas with no loss of capability, and yet avoid being detected by radar.

The first aircraft to employ stealth capability was the U.S. F-117A Nighthawk, originally codenamed Senior Trend, which had its first test flight in 1981. The F-117A is produced by Lockheed Aeronautical Systems, and the first combat-ready aircraft was delivered to the U.S. Air Force in August 1982. U.S. Air Combat Command's 4450th Tactical Group, the only F-117A unit, achieved operational capability in October 1983. The F-117A remained classified until 1988 and was first revealed to the public in 1990.

AIRCRAFT CARRIER

The first vessel specifically designed for use as an aircraft carrier was HMS *Argus*, which was launched in 1917 and used at the end of the First World War.

The *Argus*'s keel had originally been laid down as an Italian cruise liner, which would have been named the *Count Rosso* if it had been launched under the Italian flag, but, before construction was completed, the Royal Navy commandeered the ship and had it installed with a flush or flattop deck to serve as a takeoff and landing area. The flush deck became the standard configuration for future aircraft carriers.

KAMIKAZE

Ritual suicide once formed part of the Japanese samurai code for the atonement of failure or loss of face, and was regarded as an honorable act. During the Second World War, volunteer Japanese pilots known as kamikaze committed to fly their planes on suicide missions, directly into enemy ships. In a formal preparatory ritual, the pilots were bolted into their specially built aircraft. The pilots regarded their deaths as the most honorable way of serving their emperor and their country, believing, as they did, that they were the natural successors to the ancient samurai.

The first kamikaze attack took place in October 1944 against the U.S. aircraft carrier *St. Lo*, which was sunk after twenty-six kamikaze aircraft were detailed to attack.

Kamikaze translates as "divine wind," a reference to the typhoon that drove Kublai Khan's invading fleet away from the shores of Japan in 1281. The underwater remains of Khan's fleet have recently been discovered by archaeologists.

NUCLEAR WEAPONS

*The energy produced by the breaking down of the atom
is a very poor kind of thing. Anyone who expects a source
of power from the transformation of these atoms
is talking moonshine.*
—LORD RUTHERFORD (1871–1937)

Atomic fission was discovered in 1938 at the Kaiser Wilhelm Institute in Berlin by German chemist Otto Hahn (1879–1960) in cooperation with Dr. Fritz Strassman (1902–80), a radio chemist.

The first detonation of an atomic bomb, known as the Trinity Test, took place on July 16, 1945, in the desert of New Mexico. The test was a spectacular success, and eyewitnesses from twenty miles away reported feeling the heat of the explosion.

The first atomic bomb detonated in war was dropped on Hiroshima on the mainland of Japan on August 6, 1945. Ninety percent of the city was destroyed. Out of a population of around 250,000 people, about 45,000 died on the first day, with a further 19,000 dying in the next four months.

Previously, on July 25, 1945, U.S. president Harry Truman (1884–1972) issued the order to General Carl "Tooey" Spaatz (1891–1974), commander of U.S. Strategic Air Forces in the Pacific, which would result in the bombing of Hiroshima. Truman noted in his diary that he had ordered the bomb to be dropped on a "purely military target."

The Soviet Union developed its first atom bomb in 1949 following the betrayal of U.S. nuclear secrets to the Soviet Union by the spy Klaus Fuchs (1911–88).

Fuchs was born in Germany and fled to England in the 1930s when the gestapo began to round up communists. He worked on British atomic bomb research before being transferred to the United States to work on the Manhattan Project, which led to the development of the atomic bomb. After his arrest for spying in 1949, and his trial in 1950, Fuchs was sentenced to fourteen years in prison, of which he served nine. On his release, Fuchs illegally relocated to East Germany, one of the Soviet vassal states, where he began lecturing in physics.

The world's first hydrogen bomb was detonated by the United States on November 1, 1952. The explosion of the device, which was code-named Mike, caused the island of Elugelab in the Pacific to disappear, leaving a crater 1 mile wide and 160 feet deep. In the process 80 million metric tons of earth were lifted into the air. The characteristic mushroom cloud rose to 57,000 feet in ninety seconds, and eventually spread to a width of 1,000 miles. The results so terrified Norris Bradbury (1909–97), the director of Los Alamos National Laboratory, the American atomic research center, that he considered keeping the magnitude of the detonation secret.

The bomb was the eventual result of a meeting in 1949 between President Truman and Edward Teller (1908–2003), known as the father of the hydrogen bomb, who pressed for an urgent study to be undertaken for the development of a super bomb. The plan was to build a thermonuclear device, which ultimately came to be known as the hydrogen bomb. Truman authorized a crash development program, which was immediately opposed by many others, including Robert Oppenheimer (1904–67), who had led the team that developed the first atom bomb in 1945.

The first true hydrogen (fusion) bomb tested by the Soviets was detonated on November 22, 1955. Within the Soviet Union it was named Sakharov's "Third Idea," having been designed by the famous scientist, and later dissident, Andrei Sakharov (1921–89).

The first submarine-launched nuclear missile, known as Polaris, was deployed by the U.S. Navy. It was test launched on July 20, 1960, the first ever underwater rocket launch.

U.S. president John F. Kennedy (1917–63) came to an agreement with prime minister Harold Macmillan (1894–1986) to supply Britain with Polaris nuclear missiles. The Polaris Sales Agreement was signed in 1963, and Polaris missiles were installed on British submarines by the Royal Navy in 1970.

WAR CRIMES

The first war crimes tribunal was set up in Nuremberg in 1945 for the prosecution of Nazi war criminals after the end of the Second World War. The trials lasted until 1949. Several hundred prisoners were detained on suspicion of war crimes, and in the first trial, which indicted the twenty-four most important of the accused, twelve death sentences were handed down and executed. Three of the accused were acquitted.

CONCENTRATION CAMPS

One of the most reviled aspects of the Nazi regime during the Second World War was its use of concentration camps. Six million Jewish prisoners and up to 4 million others are known to have died in these camps in appalling circumstances at the hands of the Nazis.

The Japanese also built concentration camps throughout Indochina and Manchuria, and prisoners were badly mistreated and exploited for slave labor. Adding to the misery, medical experiments were carried out on prisoners in German and Japanese camps, but neither the Germans nor the Japanese were the first to make use of such camps.

The first use of concentration camps was during the Third Cuban War of Independence (1895–98), which pitted the local populace against its Spanish occupiers. In an act of desperation to stop attacks by insurgents from the countryside, the Spanish governor of Cuba, Valeriano Weyler (1838–1930), organized a mass relocation of the noncombatant rural population into specified areas within cities. He called these areas reconcentration camps, but failed to provide adequate medical treatment or sufficient food. Hundreds of thousands died, and the legacy of bitter resentment is felt to this day.

During the Boer War of 1899–1902, in South Africa, the British Army built 109 concentration camps to hold Boer women and children, and other prisoners. The camps had been built during fighting between the British and the disaffected Dutch and German settlers in the Transvaal. Forty-five of the camps were for white people and sixty-four for black people. The initial plan in London had stressed the need for humane treatment of the prisoners, but in reality the treatment was brutal. Poor food rations, inadequate hygiene, and lack of proper medical facilities led to outbreaks of typhoid and dysentery. Nearly 28,000 Boer prisoners died, of whom 22,000 were under sixteen years old. At least another 14,000 black prisoners died.

AWARDS

PURPLE HEART

The Purple Heart was the first American military award for bravery in action against the enemy and was instituted in 1782 by George Washington. In its original form, the award was a simply a piece of cloth that was edged with silver braid, and sewn into the garment. In the first year of its existence, Washington awarded three Purple Hearts, two of which still exist intact today.

The awarding of Purple Hearts fell into disuse for around 150 years, until it was revived on February 22, 1932. Today the Purple Heart is awarded to anyone wounded or killed in action against an enemy. The specification of a "wound" is laid out in U.S. Army Regulation AR 600.8.22. Providing the intention was to fire at the enemy, wounding by so-called friendly fire also qualifies. The Oak Leaf Cluster is awarded to those who have been wounded a second time, after already having been awarded the Purple Heart.

MEDAL OF HONOR

Also referred to as the Congressional Medal of Honor (because it is awarded in the name of Congress), the Medal of Honor is the highest military award of the United States of America and is one of only two decorations worn around the neck. It is awarded for valor during combat at the risk of one's own life and beyond the call of duty. The Navy version was signed into law by Abraham Lincoln in 1861 and the Army version in 1862. All branches of the armed services now qualify.

The only woman to have been awarded the Medal of Honor is Dr. Mary Edwards Walker (1832–1919). Her award was specifically for services at the first battle of Bull Run (1861), during which she continuously crossed the battle lines at great danger to herself. In 1917 the U.S. Congress changed the criteria for the award, and along with nine hundred others, Dr. Walker found her medal had been rescinded. She refused to return it and continued to wear it for the rest of her life. Her medal was restored posthumously by President Carter in 1977.

VICTORIA CROSS

The Victoria Cross is the highest honor awarded to British servicemen in the face of the enemy. It was instituted on January 21, 1856, during the Crimean War, and each medal is cast in bronze from parts of captured cannon taken at Sebastopol (now Sevastopol).

ORIGINS OF SOME WARS

ROMAN CONQUESTS

The Romans did not set out with a plan to subdue the world and introduce Roman culture into its conquered territories; each conquest seemed to lead to the next until the Roman Empire spanned most of Europe and large swathes of Africa and Asia.

In 509 BC, Rome temporarily came under the control of the Etruscans, who came from north of Rome. The expulsion of the last Roman king, the violent Tarquinius Superbus, by the Etruscan king Porsenna, signaled the beginning of the Roman republic. Porsenna left Rome before he could assume the monarchy, and at the end of the fifth century BC Rome began to take territory from the Etruscans. Rome's first major war of expansion lasted between 437 and 426 BC, and ended in victory over the town of Fidenae. The important Etruscan city of Veii was taken in 396 BC. By 275 BC, Rome controlled the whole of the Italian peninsula and then embarked on the domination of most of the known world.

ALEXANDER THE GREAT'S CONQUESTS

After he had succeeded his father, Philip II of Macedon, Alexander the Great began his conquests in 336 BC. His driving ambition was to expand the Macedonian empire, conquering Persian-dominated Asia Minor, Syria, Egypt, Persia itself, and reaching as far as India.

THE NORMAN CONQUEST

The Norman Conquest of England originated from William of Normandy's claim to the throne of England. In 1066 he defeated Harold Godwinson at the Battle of Hastings, and he became known as William the Conqueror (1028–87), and king of England.

THE CRUSADES

The First Crusade took place in 1097 out of a desire to protect the Holy Land from the Muslims. Jerusalem was captured in 1099.

THE HUNDRED YEARS' WAR

These were conflicts involving attempts by the English kings to claim the French crown and dominate France. They started around 1337 in the reign of Edward III and ended about 1453 in the reign of Henry VI, when the vast majority of the captured territory in France had been recovered by the French crown. However, Calais was not recovered until 1558.

THE THIRTY YEARS' WAR

What is known as the defenestration of Prague, in 1618, marked the beginning of the Thirty Years' War. It was a protest by Bohemian Protestants against a violation of their religious rights and led to a revolt against the Hapsburg emperor Ferdinand II. The war, which was mainly fought in central Europe, spread to involve most of the countries of Europe and was fought generally on religious lines between Protestants and Catholics. It ended with the Treaty of Westphalia in 1648.

THE ENGLISH CIVIL WAR

In 1642 the English Civil War was precipitated by the attempt by King Charles I to arrest five members of Parliament. The attempt failed, following which the Royalists, who supported Charles I, and the Parliamentarians, prepared for war.

There had been a long-running background of conflict between the two sides, with Charles insisting on his right to rule, especially to collect taxes, without interference, whereas Parliament wanted a reduction of Charles's power. Charles insisted on preserving his powers and, in 1649, after losing the war, he was tried and beheaded. Oliver Cromwell (1599–1658) forcibly dissolved Parliament and was installed as lord protector in 1653.

THE AMERICAN WAR OF INDEPENDENCE

When a British force was sent to Concord to destroy American rebel stores in 1775, the American War of Independence began. The battles of Lexington and Concord were the opening skirmishes in the war that lasted until 1783, resulting in the independence of the thirteen American colonies, which eventually became the United States.

The war was preceded by years of strife between the colonies and Britain. One of the most famous events in history, the Boston Tea Party of 1773, sparked the events that led to the war. Britain was extracting tax from the colonists without allowing them any voting rights. The watchword became, "No taxation without representation." On the night of December 16, 1773, hundreds of cases full of tea were emptied into Boston harbor (see "Food and Drink," p. 118).

THE CRIMEAN WAR

In 1853 a dispute arose between Russia, which demanded the right to protect Orthodox Christians in the Ottoman Empire, and the Ottoman sultan. There was also a dispute between Russia and France over the privileges of Orthodox and Catholic churches in the holy places in Palestine. An alliance between France, the Ottoman Empire, and Britain was formed to fight Russia, mainly on the Crimean Peninsula. The war lasted until 1856.

THE AMERICAN CIVIL WAR

It is generally supposed that the American Civil War started in 1861 over the abolition of slavery, but debate still rages over the undisputed reason for this complicated war. Twenty-three states, known as the Union, were opposed by eleven Southern states, known as the Confederacy.

The eleven states had seceded from the United States during the early stage of Lincoln's presidency, but were decisively returned into the union after the bloody war, which began with the Confederate attack on Fort Sumter in 1861. About 3 percent of the population were casualties in the war, which lasted until 1865.

BOER WAR

Actually, there were two Boer wars. The first, which is more properly known as the Transvaal War, took place in 1880–81, and was a victory for the Boers, who kept their disputed territory. The British government signed a peace treaty that allowed the Transvaal to be self-governed.

The Second, and better known, Boer War, also known as the South African War, began in 1899. The British were again in dispute with Dutch and German settlers, this time over the Orange Free State and the Transvaal territory in South Africa, following the discovery of major gold deposits around Johannesburg. The immediate cause of the war was an ultimatum made by the Boer leader, Paul Kruger, against British reinforcement of a garrison. Convinced that war was inevitable, the Boers invaded Natal Province and Cape Colony in late 1899.

Following the defeat of the Boers in 1902, both the Transvaal and the Orange Free State became part of the British Empire.

FIRST WORLD WAR

On June 28, 1914, Archduke Franz Ferdinand of Austria (1863–1914) was shot dead in Sarajevo by Serbian student Gavrilo Princip (1894–1918). This event precipitated the First World War, creating a domino effect, which triggered the implementation of international treaties following Austria's invasion of Serbia and Germany's invasion of Belgium. The treaties brought Britain, Russia, and France into the conflict.

SECOND WORLD WAR

Germany invaded Czechoslovakia in March 1939 and Poland on September 1, 1939. This second invasion led to Britain and France declaring war on Germany two days later, and established September 3, 1939, as the start of the Second World War.

Some historians have argued that the invasion of China by Japan in July 1937 was the actual start, although this is not generally accepted.

CHINESE CIVIL WAR

In 1927 Generalissimo Chiang Kai-shek (1887–1975), the leader of the Chinese Nationalist Party, purged Communists from the alliance that had been set up between the Nationalists and the Chinese Communist Party. Spanning the Agrarian Revolution of 1927–37, the fabled Long March of 1934–35, the power struggle of 1945–47, and the final struggles from 1945–50, the Chinese Civil War established Communist control of mainland China in 1950. Mao Tse-tung (1893–1976) assumed the Chinese leadership, and Chiang Kai-shek fled to Taiwan.

KOREAN WAR

The first attack of the Korean War came on June 25, 1950, when the Communist North Koreans, with Soviet backing, launched a massive attack on South Korea, across what was known as the thirty-eighth parallel. The thirty-eighth parallel was a line drawn across the map of Korea in 1945, to establish the north-south boundary.

After this attack the United Nations called on all its members to halt this aggression and the United States sent troops to defend South Korea. The conflict became a proxy war between the United States and its allies, and the Communist bloc, including China.

The Korean War established the precedent of the United States defending territories under attack from Communism, and after more than 3 million people had died, an armistice was agreed upon on July 27, 1953.

VIETNAM WAR

Similar in many ways to the Korean War, the Vietnam War was provoked by the efforts of the United States and the South Vietnamese to stop the spread of Communism from North Vietnam.

The French had tried to reestablish control over Vietnam in 1945, having lost control during the Second World War. With Vietnamese troops

(Vietminh) under the command of Ho Chi Minh, the French were badly defeated. The 1954 battle of Dien Bien Phu was decisive in French withdrawal. The United States entered the conflict by providing economic aid in 1956, and the last U.S. personnel were withdrawn in 1975.

FALKLANDS WAR

In the middle of a devastating economic crisis in Argentina, with massive civil unrest and public criticism of the ruling military junta, President General Leopoldo Galtieri (1926–2003) tried to establish sovereignty over the Falkland Islands, or Malvinas as they are called by the Argentines. His main objective was to divert attention from the problems at home.

The first offensive action of the war was the invasion of the small island of South Georgia by fifty Argentine fishermen on March 19, 1982. The invaders proceeded to raise the Argentine flag and a full-scale occupation of the Falkland Islands took place on April 2, 1982. British forces recaptured the capital, Stanley, on June 14, 1982, and by June 20 hostilities came to an end with the retaking of the South Sandwich Islands.

FIRST GULF WAR

Although Kuwait had been a British protectorate following the end of the First World War, and an independent country since 1961, the territory had been claimed by various Iraqi rulers, particularly after the discovery of oil there in 1938.

The immediate lead-up to this conflict came from Iraqi accusations that Kuwait was illegally directionally drilling for oil across the Iraq–Kuwait border. On August 2, 1990, Iraqi forces occupied Kuwait. The United States led a coalition of forces numbering over 600,000 troops, which achieved a decisive victory with little loss of life on the coalition side. The war ended on February 28, 1991, with Saddam Hussein still in power in Iraq.

SECOND GULF WAR

The aftermath of the first Gulf War saw the continuation of tensions between the United States and its Western allies with Iraq. The World Trade Center was bombed by terrorists on September 11, 2001, and the invasion of Afghanistan followed in October 2001 in an attempt to find the perpetrators. The reasons given for the invasion of Iraq included claims that Saddam Hussein was stockpiling weapons of mass destruction (WMDs) and had continually flouted numerous United Nations resolutions. The opening skirmish came on March 20, 2003. Baghdad fell on April 9, 2003.

ESPIONAGE

Black's Law Dictionary, which was first published in 1891, defines *espionage* as, "The practice of gathering, transmitting or deliberately losing secret information related to the national defence."

The earliest documented reference to espionage was around 515 BC in Sun Tzu's (544–496 BC) classic book, *The Art of War*. In this he lists the five types of spy needed by an army commander:

- Local spies—hired in the countryside
- Inside spies—subverted government officials
- Double agents—captured enemy spies who have been "turned"
- Doomed spies—deceived professionals who take army orders to the enemy
- Surviving spies—who return with reports

The ancient Egyptians (c. 3000–343 BC) also used a well-established network of spies for gathering intelligence on potential enemies both within the state and outside.

UNITED STATES

The Central Intelligence Agency (CIA) The U.S. government's principal intelligence and counterintelligence agency was created in 1947 by President Truman as the successor to the Office of Strategic Studies (OSS). The OSS had been the chief intelligence-gathering arm of the U.S. government during the Second World War and was disbanded in 1945.

The Federal Bureau of Investigation (FBI) was established as the Bureau of Investigation in 1908 by Attorney General Charles Bonaparte (1851–1921). The name was changed to the FBI in 1935. Most of its existence has been spent in federal law enforcement, investigating the activities of political activists, most especially during the period of the cold war from the 1940s through to the 1970s.

Note: Charles Bonaparte was the grandson of French emperor Napoléon Bonaparte's younger brother, Jérôme.

BRITAIN

Britain's Intelligence Service was set up along modern lines under Elizabeth I (1533–1603) and this long history has been instrumental in the structure of other intelligence services.

MI6 was founded in 1909 by Sir George Mansfield Smith-Cumming (1859–1923) as the overseas arm of the Secret Intelligence Service (SIS). He began the practice of signing his letters and memos in green ink with the letter C. This practice was adopted by all subsequent holders of the office of director of SIS, hence they were all referred to as "C." This was also the idea behind James Bond's chief being known as M.

MI5 was founded in 1909 as the intelligence agency responsible for internal security and domestic counterintelligence. It was originally designated the Directorate of Military Operations Section 5 (MO5). The first director general was Sir Vernon George Waldegrave Kell (1873–1942), who doubled as foreign correspondent for the *Daily Telegraph*.

The first female director general of MI5 was Stella Rimington (now Dame) (b. 1935) who served between 1992 and 1996.

SOVIET UNION

The **KGB** (committee for state security) is the best known of the old Soviet state security services. It was formed in 1954 as successor to the infamous NKVD, and operating through many offices has exercised pervasive influence on the people of the former Soviet Union and those countries and organizations with whom the Soviet Union had any contact. With the dissolution of the Soviet Union in 1991, it came under the control of the Russian government with greatly reduced powers and influence.

The **NKVD**, headed by Lavrenty Beria (1899–1953), was the forerunner of the KGB and was itself a descendant of the Cheka, which was set up in 1917 by Polish-born Felix Dzerzhinsky (1877–1926), one of the founding fathers of the Russian Revolution.

CHINA

Chinese intelligence is thought to be organized along the same lines as that of the old Soviet Union. It is also thought that the whole structure was abolished during the Cultural Revolution in the 1960s and reformed with the military taking control of the civilian intelligence agencies.

ISRAEL

Mossad, more properly known as the Central Institute for Intelligence and Security, was formed immediately after the creation of the state of Israel in 1948. Mossad's duties include foreign espionage and the conducting of covert political operations. Israeli agents living in Arab lands come under the control of Mossad.

GERMANY

The gestapo (Geheime Staatspolizei) was established in 1933, taking over the role of the Prussian secret police. Hermann Göring (1893–1946) took charge from 1934, expanding the gestapo's role over the whole of Germany with the exception of Bavaria, which came under Heinrich Himmler (1900–45) and his SS troops. Later that year Göring handed over control of the gestapo to Himmler.

CODE BREAKERS AND CODE BREAKING

Far from the supposedly "glamorous" end of spying, sit the code breakers, the men and women whose job it is to decrypt enemy messages.

The first treatise on code breaking, *On Deciphering Cryptographic Messages*, was written in the ninth century by the Arab scientist and mathematician Al-Kindi (AD 801–73) who lived in what is present-day Yemen.

Perhaps the most celebrated code breakers were the team at Bletchley Park during the Second World War, who were led by Alan Turing (see "Inventions," p. 148). They succeeded in breaking the seemingly unbreakable German Enigma code using the famous Turing-Welchman Bombe computers, the first time electronics had been used in code breaking.

The U.S. code-breaking operation within the State Department— MI8—was closed down in 1929 on the orders of Secretary of State Henry Stimson (1867–1950), with the words, "gentlemen do not read each other's mail."

FEMALE SPIES

The first known British female spy was Ann Bates, a Philadelphia schoolteacher, who spied for the British and penetrated George Washington's inner circle. During the years 1778–80, Bates disguised herself as a peddler and walked almost unchallenged through the American lines, checking gun emplacements, numbers of troops, and types of weaponry. After the American War of Independence, Bates sailed to England. She was abandoned by her husband but succeeded in obtaining a pension for her espionage work.

Pauline Cushman (1833–1893) was an actress from New York and the first female Union spy of the American Civil War. Penetrating the Confederate lines and gathering information was her first assignment. She used her acting skills to pass through the lines dressed as a man and gained access to the battle plans of General Braxton Bragg. Cushman was caught and convicted of spying but again used her acting skills to feign serious illness and to postpone the death sentence that had been passed. She kept up the act long enough to be rescued by advancing Yankees and was promoted to the rank of major by Abraham Lincoln.

SPY PLANES

The first spy plane was the 1917 De Havilland DH4, equipped with a mounted camera, which was fixed to the fuselage for vertical photography. Vibration proved a drawback until the revolutionary K-3 camera, which was developed in 1920 by the American inventor and entrepreneur Sherman Fairchild (1896–1971). Fairchild was later to found Fairchild Semiconductor, which made major contributions to the advancement of computers.

FICTIONAL SPIES

James Bond was the brainchild of Ian Fleming (1908–64) and was introduced in *Casino Royale*, published in 1953.

George Smiley first appeared in *Tinker, Tailor, Soldier, Spy*, a novel by John Le Carré (b. 1931), which was first published in 1974. Smiley appeared in two other Le Carré books, *The Honourable Schoolboy*, in which he was not the main character, and *Smiley's People*.

TERRORISM

Those who make peaceful revolution impossible,
make violent revolution inevitable.
—JOHN F. KENNEDY (1917–63)

In the *Oxford English Dictionary*, terrorism is defined as "The unofficial or unauthorised use of violence and intimidation in the pursuit of political aims."

The word *terrorism* first came into use during the French Revolution of 1789–99, when the Reign of Terror, which lasted from June 1793 through July 1794, was at its height. The Terror was effectively state terrorism: the ruling Jacobin faction led by Robespierre ruthlessly executed anyone thought to be a threat to the regime.

THE ZEALOTS

The earliest example of terrorism appears to be the campaign waged by the Zealots, a Jewish political sect, against the Roman occupiers of Israel during the first century. Zealots were known as "dagger men" as they frequented public places with daggers concealed beneath cloaks. Without warning, they would strike down people known to support Rome.

The Zealots conducted an unrelenting anti-Roman campaign in the eastern Mediterranean region during the Jewish revolt of AD 66–70, managing to capture Jerusalem at one point. Masada, a mountain fortress in southern Israel, was the site of the Zealots' last stand in AD 70 after Jerusalem had been lost. The Romans conducted a two-year campaign, and when it was inevitable that Masada would be captured, the Zealots committed mass suicide, with only seven women and children surviving.

The Jewish historian Josephus (circa AD 37–100) tells of their murderous activities, and they were criticized in the Talmud.

IRISH REPUBLICAN ARMY

Many regard the 1916 Proclamation of the Republic, made during the Easter Rising, as the founding document of the Irish Republican Army (IRA). There had been a history of armed uprisings against British rule well before the proclamation, with the most notable ones taking place in 1798, 1803, 1848, and 1867. (The Irish Republican Brotherhood, supported by American money, carried out the 1867 attacks.)

The Ulster Volunteer Force (UVF) was formed in 1913 and in 1914 was allowed by the British to import arms, unhindered. Until the Easter Rising of 1916, the UVF had been regarded as little more than "toy soldiers," according to IRA leader Ernie O'Malley (1897–1957), speaking in 1923, but by 1917 the UVF had managed to achieve strong support throughout Ireland and were regarded as a viable fighting force.

The UVF changed its name to the IRA in 1919, and began its guerrilla campaign. Bombings of military and civilian targets became the tactic of choice. These tactics, employed effectively by such legendary divisions as Tom Barry's Flying Column in Cork, which made large parts of Ireland ungovernable, became textbook examples of this type of armed struggle and became the inspiration for other similar organizations throughout the world.

Tom Barry (1897–1980) who had previously enlisted in the British Army, regarded the Crown forces of 1920 as the real terrorists.

AL-QAEDA

To most people in the West, al-Qaeda is seen as a byword for terrorism, a single, highly centralized structure, with Osama bin Laden as its shadowy mastermind and leader. However, it is now widely believed that al-Qaeda does not exist in this form and is in fact merely a convenient label applied to the general groupings of Islamic militants (or Islamists). The literal meaning of *al-Qaeda* can be "base" or "foundation." Alternatively, it can mean "rule" or "maxim."

Islamic militancy in its present form sprang from the Soviet occupation of Afghanistan in 1980, and the rise of the Mujahedeen resistance fighters. The stated aims of the militant Islamic groups, commonly referred to as al-Qaeda, are to overthrow secular Arab regimes and to reinstate the caliphate across the Arab world. Indiscriminate suicide bombing of civilians, at home and abroad, is the tactic of choice.

HEZBOLLAH

Hezbollah is an Arabic word that means "the party of God."

Hezbollah is an Islamist extremist organization based in Lebanon. There are conflicting accounts of the formation of Hezbollah, with some sources listing 1982 as the official foundation date and others claiming it as February 16, 1985. It was founded by Sheikh Ibrahim al-Amin with the original objectives of bringing Islamic revolution to Lebanon and the destruction of the state of Israel. Today it is a well-structured political party with its own TV station, a monthly magazine, and several Web sites, including some for recruitment.

Although Hezbollah has publicly condemned the bombing of the twin towers in 2001 and the targeting of civilians, the organization pioneered the use of suicide bombers. It is estimated that Hezbollah has taken delivery of up to 15,000 Katyusha rockets and around 30 Zelzal-2 unguided missiles (nicknamed "earthquake bombs"), which have the capability of delivering a 1,300-pound warhead a distance of 130 miles. Iran claims to have supplied the weapons.

HAMAS

An Islamist paramilitary organization based in Palestine, Hamas (Harakat al-Muqawamah al-Islamiyyah) has evolved through a number of stages: The two most important of these are the founding of the Muslim Brotherhood in the Gaza Strip between 1967 and 1976, and the formation of Hamas as the combatant arm of the Muslim Brotherhood in 1982.

The stated aims of Hamas are to eliminate the state of Israel, and to create an Islamic theocratic state on the land currently in the possession of Israel, including the West Bank and Gaza Strip. Suicide bombings have been the main tactic of choice, including the use of female suicide bombers.

In February 2006, Hamas won an overwhelming success at the polls and was elected the ruling party of the Palestinians.

BRIGATE ROSSE

Renato Curcio (b. 1945) founded the Brigate Rosse (Red Brigades), an extreme Italian Marxist/Leninist group, in 1969 while he was still a university student. The stated aim was to create a revolutionary state through armed struggle, and their subsidiary objective was to take Italy out of the Western alliance.

In November 1970 Brigate Rosse announced their existence with fire-bombings in Milan. Another preferred tactic is assassination of government ministers and business leaders, and in 1978 they captured the prime minister of Italy, Aldo Moro (1916–78), and murdered him.

By the end of the 1980s, following the imprisonment of many of its leaders, the organization was greatly weakened and remains almost ineffective.

THE SHINING PATH

Abimael Guzman (b. 1934), a university philosophy professor, founded the Shining Path in the late 1960s, as an offshoot of the Communist Party of Peru. Shining Path's stated aim is to replace Peruvian bourgeois institutions with a Communist peasant revolutionary regime. Guzman was captured and jailed in 1992, since when Shining Path has not been active.

Between 1973 and 1975, Shining Path gained control of student councils and adopted a Maoist "criticism and self-criticism" doctrine, which led to students denouncing their peers as being insufficiently revolutionary. Their savagery against union leaders and peasants has led to widespread condemnation.

EUZKADI TA AZKATASUNA

Young nationalists seeking independence for the Basque region from Spain founded Euzkadi Ta Azkatasuna (ETA) in 1959.

ETA is an armed organization, which considers the unique Basque language, Euskera, to be a national defining characteristic. It aims, by violent means, such as assassination and murder, to create a Basque state separate from Spain. The earliest death is reported to have been a baby killed in a June 1960 bombing attack.

Shootings and bombings are the tactics of choice, and despite much lower levels of activity since the 1997 murder of a twenty-nine-year-old local councillor, more than eight hunded people have been killed by ETA since its formation. However, it is now thought that ETA no longer believes it can achieve its aims by violent means.

AUM SHINRIKYO

On March 20, 1995, sarin gas was released on the Tokyo subway, resulting in the deaths of twelve people. The attack was by far the most widely reported action of Aum Shinrikyo (renamed Aleph), a controversial neoreligious group led by Shoko Asahara (b. 1955). Aum claims to be guided by Buddhist and Hindu doctrines.

Shoko has been sentenced to death by hanging.

JAPANESE RED ARMY

Fusako Shigenobu (b. 1945) founded the Japanese Red Army (JRA) in 1971 as a breakaway group from the Japanese Communist League. The stated aims of the JRA were to overthrow the Japanese government and monarchy and to start a world revolution. During her travels around Europe and the Middle East, Fusako forged strong links with the PFLP

(Popular Front for the Liberation of Palestine), which gave help with training, finance, and weapons.

Fusako was arrested in Osaka in 2000 and sentenced to twenty years in prison.

KHMER ROUGE

Khmer Rouge was formed at the 1959 Congress of the People's Revolutionary Party of Cambodia, but kept its name secret until 1967.

Under the leadership of Brother No. 1, Pol Pot, born Saloth Sar (1925–98), Khmer Rouge launched national insurgency and a guerrilla war across Cambodia in 1968. It managed to gain control of the country in 1975 and continued to rule until 1979. Estimates of the deaths directly attributed to Khmer Rouge during this period vary between 1.7 million and 3.3 million.

Pol Pot "officially" dissolved the organization in 1996, but a few of the surviving leadership await trial for crimes against humanity.

MISCELLANEOUS

BULLETPROOF VEST

The first U.S. patent for a bulletproof vest was issued in 1919, but body armor was not introduced into general police use until 1931, when the vest's effectiveness was demonstrated to the Washington police department.

The so-called bulletproof vest was not successful against high velocity bullets until Kevlar was introduced in 1971. Kevlar is a lightweight, high-strength synthetic polymer fiber, which was the first to offer the level of protection needed for stopping bullets. It was developed in 1965 in the laboratories of the American company DuPont, by Stephanie Kwolek (b. 1923).

PEACE TREATY

The earliest known peace treaty was concluded in 1269 BC between the Egyptian pharaoh Rameses II and King Hattusilis of the Hittites, following the battle of Kadesh, in modern day Syria, in 1275 BC. The battle was inconclusive, although Rameses claimed it as a great triumph.

The language of the treaty is Akkadian, and it is written in cuneiform script on clay tablets, fragments of which still exist. The greater threat to both countries by invaders known as the "sea peoples" encouraged the signing of the Treaty of Kadesh, which covered nonaggression and mutual protection in the event of invasion of one country by an enemy.

ZOOS

COVERING ZOOS

Modern zoos are places of education, conservation, and research. In the Western world they are subject to the scrutiny of animal welfare organizations, are staffed by dedicated, professional animal experts, and are generally open to the public. Zoos were not always this way: they began as places of entertainment for the privileged few.

The first known example of birds held in captivity is in 4500 BC, in Arpachiya (modern-day Iraq). Pigeons were kept captive in large numbers for exhibition.

The earliest examples of animals held in captivity come from both India and Egypt around 2500 BC. The Mohenjo Daro civilization of India kept elephants for both work and exhibition, and the ancient Egyptians kept a variety of exotic animals, including lions, mongooses, and baboons, which were generally reserved for the amusement of royalty.

Two great zoos were constructed in ancient China In 1150 BC the Empress Tanki had a "house of deer" built of marble, and in around 1000 BC, the Emperor Wen Wang established what he called the Garden of Intelligence for the housing of rare animals. The grounds of the Garden of Intelligence extended to 1,500 acres.

King Solomon (reigned 962–22 BC), who is widely regarded as the greatest of the ancient kings of Israel, not only established a massive harem, but also a zoo, in about 930 BC.

The first serious study of animals began in the zoos of ancient

Greece, between 700 and 200 BC. In 340 BC, after detailed research in the zoos, the Greek philosopher Aristotle (384–22 BC) wrote his book *The History of Animals*. Aristotle's most famous pupil, Alexander the Great (356–23 BC), arranged for the capture of strange and exotic animals on his military campaigns, which he had shipped back to Greece for study.

The Greek philosophers Epicurus (341–270 BC) and Anaximander (circa 610–546 BC) both developed ideas about evolution. These were not fully developed until Charles Darwin's work in the nineteenth century.

The first to stage large-scale fights between animals, and later between animals and gladiators, were the ancient Romans. The animals had initially been kept for observation and display but were quickly used for public entertainment, including the spearing of animals by the audience.

It is thought that modern zookeeping began in 1752 with the opening of the zoo at the Tiergarten Schönbrunn in Vienna. The Empress Maria Theresia (1717–80) and her husband Franz Stephan (1708–65), who later became Holy Roman Emperor Francis I, commissioned the design and construction of the zoo at their summer palace at Schönbrunn. The zoo was opened to the public in 1778 but only to "decently dressed persons," and only on Sundays.

The term *zoo* was first used in 1826 after the foundation of the Zoological Society of London. The principal founders were Sir Stamford Raffles (1781–1826), who also founded Singapore in 1819, and the physicist Sir Humphry Davy (1778–1829) who, among many other things, invented the miners' safety lamp, which bears his name. The stated objective of the Zoological Society was "the advancement of zoology and animal physiology, and the introduction of new and curious subjects of the animal kingdom."

The Zoological Society also led the way in publishing the first scientific journal on the study of animals, and its *Journal of Zoology* has been continuously published since 1830. A Royal charter was granted to the Zoological Society by George IV (1762–1830) on March 29, 1829.

The first zoo in the United States was New York City's Central Park Zoo, which opened in 1864. In the four years leading up to the authorization of the zoo by the state legislature, the growing collection of animals was put together through donations from members of the public. A black bear cub and seventy-two white swans were among the donations.

Although grizzly bears, captured by the famous bear hunter James "Grizzly" Adams (1812–60), had been kept for exhibition in a San Francisco basement as early as 1856, the San Francisco Zoo, known as Woodward's Gardens, was not opened until 1866.

The London Zoo (Zoological Society of London) opened its collection to the public in 1847, although it had originally begun to collect the animals for scientific research in 1828. Sited on the north side of Regent's Park, London Zoo claimed to be the world's first scientific zoo (ignoring Aristotle's work in the fourth century BC). The full title of the collection of animals was the Zoological Gardens, but this was soon shortened to "the zoo" by the visiting public, and the word *zoo* passed into common use.

The revolutionary breakthrough in zoological studies came on November 24, 1859, with the publication of *On the Origin of Species by Means of Natural Selection*, written by Charles Darwin (1809–82). In his book, Darwin put forward his now generally accepted theory of evolution, in which traits arising from adaptation are passed from generation to generation.

The original paper on the evolution of species was presented to the Linnean Society of London on July 1, 1858. The paper was a joint presentation between Darwin and Alfred Russel Wallace (1823–1913). Darwin had spent many years formulating his theory, and Wallace had written his own paper, exactly mirroring Darwin's theory, while working in Malaya. In all innocence, he sent his findings to Darwin for advice, and it looked for a while as if he had beaten Darwin to the announcement of the theory of evolution. In the end, Wallace was persuaded by Darwin's friends to share the honors and the two became lifelong friends.

Although the theory of evolution, or natural selection, is referred to as Darwinian theory, Charles Darwin was not the first to propose the idea in modern times. His own grandfather, Erasmus Darwin (1731–1802), wrote:

> Would it be too bold to imagine that in the great length of time since the earth began ... would it be too bold to imagine that all warm-blooded animals have arisen from one living filament ... possessing the faculty of continuing to improve by its own inherent activity, and delivering these improvements by generation, world without end?

Earlier, several French botanists had proposed the basis of organic evolution. In 1815 one of their number, Jean-Baptiste Lamarck (1744–1829), had even produced an evolutionary diagram charting the progress of humanity from a single cell to man. He proposed the theory that organisms possess an "inner feeling" toward perfecting their species, thus preparing the way for evolution theory. However, none of the French theorists presented persuasive evidence of the process.

EPILOGUE

In compiling *The Book of Origins* much has been left by the wayside. I excuse myself that even the *Encyclopaedia Britannica* leaves some things out, although to be fair—not much. Let's hope what is in this book provides an insight into the origins of some important things and that it helps, if only in a pub quiz.

A good book has no ending.
—R. D. CUMMING (B. 1871)

The author and editors have made painstaking efforts to ensure that the information in *The Book of Origins* is accurate and in accordance with the latest information available. The most revered textbooks have been systematically scoured, and the World Wide Web has been googled in the quest for accuracy and precision. However, it is acknowledged that some of the facts within this book are the subject of debate.

We would be pleased to hear from readers who wish to add to the knowledge within this book. The author and publishers will be happy to include any corrections or fresh information in the next edition.

Please send all comments or suggestions to Trevor Homer either by e-mail to Plume.marketing@us.penguingroup.com or by post, c/o Plume Books, 375 Hudson Street, New York, NY 10014.

INDEX